Freeway Operations; Regional Systems Management and Operations; Managed Lanes 2018

Transportation Research Record: Journal of the Transportation Research Board Volume 2672, Issue 14

ISBN: 978-1-5108-7969-0

Printed from e-media with permission by:

Curran Associates, Inc.
57 Morehouse Lane
Red Hook, NY 12571

CURRAN ASSOCIATES INC. proceedings.com

Some format issues inherent in the e-media version may also appear in this print version.

Published by Sage Publications USA

ISBN (Print) 978-1-5108-7969-0
ISBN 2018 Printed Set (All Issues) 978-1-5108-7735-1

Printed by Curran Associates, Inc. (2019)

Additional copies of this publication are available from:

Curran Associates, Inc.
57 Morehouse Lane
Red Hook, NY 12571 USA
Phone: 845-758-0400
Fax: 845-758-2633
Email: curran@proceedings.com
Web: www.proceedings.com

Volume 2672 Number 14 December 2018

TRR
JOURNAL OF THE TRANSPORTATION RESEARCH BOARD

Contents

Freeway Operations; Regional Systems Management and Operations; Managed Lanes 2018

Articles

Article

TRR
JOURNAL OF THE TRANSPORTATION RESEARCH BOARD

Transportation Research Record
2018, Vol. 2672(14) 1–9

DOI: 10.1177/0361198118782759
journals.sagepub.com/home/trr
SAGE

Analysis of Toll Lanes Access Violation Behavior: Example of Express Toll Lanes on I-405

Hiroyuki Hinohara[1] and Yinhai Wang[1]

Abstract

Washington State has introduced express toll lanes (ETLs) on I-405 as a managed lane project. One of problems in the operation of ETLs operation is the access violation in which vehicles cross the double white lines, which separate ETLs from general purpose lanes. The objective of this paper is to identify access violation patterns and quantitatively describe the relationship between access violation frequency and traffic condition, user type and road segment characteristics. Finally, this paper develops negative binomial models to predict access violation frequency. The analysis results show that that traffic conditions factors influence violation behavior differently depending on whether the violation occurred when entering or exiting ETLs, which suggests that entry and exit violation behaviors should be analyzed separately. In addition, the finding that casual users commit violations more frequently than frequent users implies the possibility of unintentional violations. It is expected that these results support enforcement against violations and improve ETLs operation in the future.

Managed lanes are getting attention around the United States as a technique to improve efficiency on freeway facilities and manage traffic demand. The Federal Highway Administration clarified that in major metropolitan cities where drivers are constantly facing serious traffic congestion, it is preferable to improve traffic flow on existing freeway facilities rather than adding another lane, considering the lack of funding and difficulty in getting construction right-of-way (*1*).

The Washington State Department of Transportation (WSDOT) started to build a "core high-occupancy vehicle (HOV) system" in the early 1990s, with the aim of constructing 320 lane miles of HOV lanes (*2*). Some 235 miles of the HOV system had been opened as of June 2017; however, serious congestion on some HOV lanes during peak hour in Seattle has limited their travel speed benefits compared with general purpose (GP) lanes. As a result, in 2008, HOV lanes were converted to high-occupancy toll (HOT) lanes on SR167 as a pilot project to improve traffic flow using efficiently managed carpool lanes. In addition, express toll lanes (ETLs) on I405 were opened in 2015 to maximize highway productivity. There are plans to extend toll lanes from SR167 to I405 in the future.

One of the biggest problems in the operation of ETLs on I405 is illegal toll lane use. ETLs on I-405 have limited access and egress points and it is a violation for vehicles to cross the double white lines which separate ETLs and GP lanes. The Washington State Patrol (WSP) reported that 43 drivers out of 58 drivers stopped for toll lane violations on the I-405 ETLs during the first two days of operation had crossed the white double lines (*3*). In addition, PRR, Inc.'s survey of I405 users revealed concerns and complaints about safety problems caused from entry and exit violations (*4*).

The Transportation 2040 long-range transportation plan calls for a fully tolled highway system in the Puget Sound region by 2030 (*5*). It is expected that Washington State will implement more tolled projects to fund new facilities and maximize highway productivity for the future. Therefore, it is important to understand access violation behaviors in order to implement safer and comfortable toll lane projects in the future.

This paper focuses on drivers crossing double white lines on I-405 ETLs in 2016. With 2016 toll transaction data, the violation patterns of crossing double white lines are quantitatively assessed by examining the effect of traffic conditions and road design factors. This paper

[1]Department of Civil and Engineering, University of Washington, Seattle, WA

Corresponding Author:
Address correspondence to Hiroyuki Hinohara: hinohara@uw.edu

consists of six parts. A literature review summarizes previous works on double white line violations and clarifies the objective and position of this research. The Study Site and Data section describes the target road and toll transaction data, and the data processing and a negative binomial model are explained in the methodology section. The relation of access violation and traffic condition or road design factors is discussed in the result section, and the paper concludes with a summary section.

Literature Review and Objective

Previous work about behavior of drivers illegally crossing white double lines between toll lanes and GP lanes is limited. Relevant research includes simulation analysis of the impact of HOT lane access violations on traffic flow and the analysis of violation patterns by video detection systems.

Choe et al. simulated the effect of violation in HOT lane access on concurrent traffic flow (*6*). In the study, access violations are divided into two types: violations where vehicles freely move between the HOT and GP lanes disregarding the buffers, and violations where vehicles cross the buffer just before and after a tolling facility to avoid toll payment. The result from microscopic traffic simulation analysis indicated a significant effect on HOT and GP lane travel times when the violation rate is over 10%.

Jonathon and Hallenbeck examined locations and operating conditions which influence double line crossing violations on SR167 HOT lanes in Washington State using camera evaluation (*7*). The study found that violation rate increased dramatically when the GP lanes are congested. In particular, violations were frequently found at the end of toll lanes, upstream of an access point where large volumes enter a traffic stream from an on-ramp, and near the upstream end of any queue.

The objective of this research is to quantitatively identify access violation patterns and analyze the effect of traffic condition, road design factors, and user type on violation behavior, which have not been discussed in the previous research. Finally, this research develops a statistical model to predict the frequency of violations with traffic condition and road segment characteristics as explanatory variables. It is expected that the results from this research will provide important information about violation behavior in toll lane use, support enforcement against violations, and improve ETLs operation.

Study Site and Data

Study Site

The ETLs on I-405 are the target corridor in this study. The ETLs, opened in September 2015, connect NE 6th St in Bellevue to I-5 in Lynwood with a total length of 17 miles. The ETLs consist of two lanes from NE 6th St in Bellevue to around NE 160th St in Woodinville and one lane from SR522 to I-5 in Lynwood (Figure 1). ETLs are open to all vehicles on weekends, holidays, and from 7 p.m. to 5 a.m. on weekdays for free. Between 5 a.m. and 7 p.m. on week days the ETLs use a dynamic toll rate strategy, with toll prices varying from 0.75 dollars to 10 dollars based on real-time traffic conditions. Free toll rates apply to vehicles satisfying carpool requirements, which are more than three people in the vehicle between 5 and 9 a.m. and 3 and 7 p.m. on weekdays, and more than two people in the vehicle during 9 a.m. to 3 p.m. on weekdays.

Access to ETLs is strategically controlled, and there are five points in the northbound direction and seven points in the southbound direction for vehicles on mainstream traffic to enter or exit the ETLs (Figure 1). In addition, direct access points on two local roads allow vehicles to directly enter into the ETLs or get off onto local roads (NE 6th St and NE 128th St). Other road segments are restricted to access by double white lines. In Washington State, enforcement for access violations is conducted through a visual observation on site by WSP troopers. Patrol troopers will pull over the vehicles crossing double white lines and give a citation with $136 ticket for them.

Data

This research mainly utilizes 2016 toll transaction records (from January 1st in 2016 to December 31st in 2016), from 5 a.m. and 7 p.m. on weekdays when the ETLs are tolled with the dynamic tolling strategy.

The toll transaction data were provided by the WSDOT toll division, and cleaned by removing flawed and missing data. The data were collected at 21 toll plazas, which are divided into 10 plazas (N1–N10) in the northbound direction and the 11 plazas (S1–S11) in the southbound direction. The toll plazas are located right after access points on the main traffic stream, at the end points of ETLs, or at direct access points for local roads. The locations of the toll plazas for data collection are shown in Figure 1.

The variables included in the data are trip ID, date, trip number, fare, transaction type, and plaza. Noteworthy is the transaction type that shows how the vehicle was detected at the toll plaza with three categories: HOV, automated vehicle identification (AVI) and image processing (IMG). HOV and AVI indicate that the vehicle was detected using transponders such as the "Good To GO" flex pass or sticker, which requires users to create "Good To GO" online accounts and obtain a transponder in advance of lane use. Carpooling vehicles need to have a switchable pass and declare that they meet occupancy requirements by switching the pass to HOV mode so that they are detected as HOV, allowing

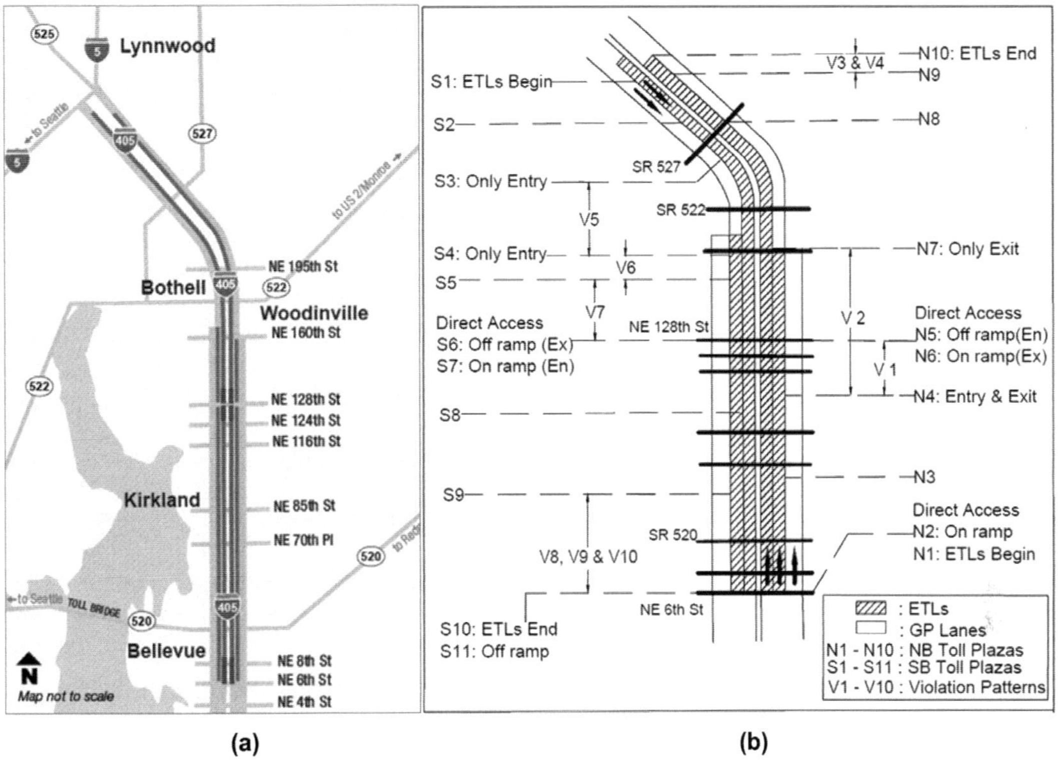

Figure 1. (*a*) Study site location (*8*) and (*b*) access points and segments with violations on ETLs.

for free ETL use. Motorcycles with motorcycle transponders are identified as HOV and also allowed to use ETLs for free. AVI includes the vehicles which have valid transponders but do not satisfy the HOV requirement and are required to pay toll fare through an online account. IMG indicates that vehicles have an invalid transponder or that the vehicle was detected by its license plate without a transponder.

Methodology

The analysis consists of two phases. First, access violation patterns are identified based on the data collection characteristics, and the relationship between access violations and road design or traffic condition is examined. In the second phase, models are developed to estimate violation frequency.

Data Processing

The focus of this research is ETL violations when vehicles illegally cross white double lines to enter or exit ETLs. Each trip ID in the data includes the toll plazas the vehicle drove through, meaning a normal trip ID will show the proper sequence of plazas. Access violations are identified by finding trips with incomplete sequence of plazas. For example, a trip with exit plaza N5 (exit at off-ramp to NE 128th St) without plaza N4 for entry means that the vehicle crossed double lines to enter into the ETLs after passing plaza N4 on the GP lanes, then exited at NE 128th St. This trip is categorized as an entry violation. Likewise, trips identified as crossing double lines to get in the GP lanes from the ETLs are categorized as exit violations. In total, 10 violation patterns are identified in this way, which are summarized in Table 1. The segments where the violations (V1–V10) occurred are shown in Figure 1. This research focuses on these 10 violation patterns and excludes other violations that would have occurred in other segments of ETLs, which cannot be identified from the data. Another limitation is that the data will exclude serious toll evasion violations which avoid paying toll altogether, because those vehicles will strategically evade detection at the toll plazas. It is assumed that the violations discussed in this study are access violations to improve driving environment or for access to local roads, rather than toll evasion violations. With this reason, the impact of toll rate on access violations is not discussed in this paper.

Table 1. Violation Patterns

Violations	Trip pattern (plaza)	Violation category	Direction	Supplementary explanation
V1	N5 without N4	Entry	North	Off ramp to 128th St
V2	N7 without N4 or N6	Entry	North	
V3	N10 without N9	Entry	North	End of ETLs
V4	N9 without N10	Exit	North	Jump in GP lanes at the ETLs end
V5	S3 without S4	Exit	South	
V6	S4 without S5	Exit	South	
V7	S6 without S5	Entry	South	Off ramp to 128th St
V8	S9 without S10 or S11	Exit	South	Jump in GP lanes at the ETLs end
V9	S10 without S9	Entry	South	End of ETLs
V10	S11 without S9	Entry	South	Off ramp to 6th St

Note: The violation numbers and plaza numbers correspond with the numbers in Figure 1.

Table 2. Summary of the Integrated Processed Data

Variables	Description
Objective variables	
Violation trips count	Frequency of trips with violations Analysis unit: 30-minute time stamp by the user type and by the violation pattern Sample size: 154,754
User segments	
User type	0: Frequent users (AVI and HOV), 1: Casual users (IMG)
Traffic conditions	
GP lanes traffic flow rate (veh/hr/ln)	Average traffic flow rate per GP lane in the same segment
GP lanes speed (mi/hr)	Average time mean speed of trips in GP lanes in the same segment
ETLs traffic flow rate (veh/hr/ln)	Average traffic flow rate per ETLs lane in the same segment
ETLs speed (mi/hr)	Average time mean speed of trips in ETLs in the same segment
Road segment characteristics	
The number of ETLs	The number of lanes in the segment
The length of segment (mi)	The length of segment where the violation occurred
Toll end dummy	0: not end, 1: the segment located at the end of ETLs
NB dummy	0: southbound, 1: northbound
Local access dummy	0: no local access, 1: the segment includes the local road access
Violation type	
Entry dummy	0: exit violation, 1: entry violation

The identified violations are converted to frequency (count) data using 30-minute time stamps and categorized by user type and violation pattern. Traffic condition variables such as the flow rate and speed on GP lanes and ETLs, which come from loop-detector data, are combined to the count data based on the segments where the violation occurred. The time mean speed is employed owing to data limitation. The traffic flow rate and speed are calculated as values per lane only on through lanes, excluding auxiliary lanes. If there are more than two loop-detector sets along the segment, average flow rate and speed are used as representative values along the segment. The processed data are summarized in Table 2. With the processed data, the impact of traffic condition, user type, and road segment characteristics on access violations is examined.

Negative Binomial Model Analysis

The final objective of this research is to develop models to predict the frequency of ETL access violations with traffic condition and road segment characteristics as explanatory variables. In this research, the processed dataset is divided into entry violation count data and exit violation count data, and two models are developed separately.

Generalized linear models such as the Poisson regression model and the negative binomial (NB) regression model have been widely applied to count data. In the Poisson regression model, the mean, λ, of the objective variable is assumed to be constant. The NB model assumes that λ follows a Gamma distribution and accommodates for over-dispersion when the objective variable shows a larger variance than the mean.

Table 3. Result of Dispersion Test

Model	Alpha ($\propto$)	z	p-value
Entry violation model	0.449	8.19	2.2E-16
Exit violation model	0.756	9.30	2.2E-16

Whether the objective variable is over-dispersed is checked through the dispersion test with dispersion parameter $\propto$. The null hypothesis assessed in this test is $\propto = 0$. When $\propto = 0$, $E(Y_{ijk}) = Var(Y_{ijk}) = \lambda_{ijk}$ and the Poisson model is appropriate. On the other hand, when $\propto > 0$, the objective variable Y_{ijk} has over-dispersion and it is concluded that the NB model is the appropriate model. Table 3 shows the result of the dispersion test for the entry violation frequency dataset and the exit violation frequency dataset, which shows that alpha $\propto$ is significantly greater than 0 for both datasets. As a result, the NB model is applied to the entry and exit violation datasets.

The NB regression model and dispersion parameter $\propto$ used are shown in the equations below:

$$Y_{ijk} \sim negative\ binomial\ distribution(\lambda_{ijk}, \theta)$$

$$\Pr(Y_{ijk} = y_{ijk}) = \frac{\Gamma(y_{ijk} + \theta)}{\Gamma(y_{ijk} + 1)\Gamma(\theta)} \left[\frac{\theta}{\theta + \lambda_{ijk}}\right]^{\theta} \left[\frac{\lambda_{ijk}}{\theta + \lambda_{ijk}}\right]^{y_{ijk}}$$

$$y_{ijk} = 0, 1, 2, \ldots$$

$$\ln(\lambda_{ijk}) = \beta X_{ijk}$$

$$E(Y_{ijk}) = \lambda_{ijk}$$

$$Var(Y_{ijk}) = \lambda_{ijk} + \lambda_{ijk}{}^2\theta^{-1}$$

$$\propto = \theta^{-1}$$

where:

Y_{ijk}: *Frequency of access violations by the observation ijk*

i: 30 *minute time stamp*

j: *Violation pattern*(10 *violation patterns*)

k: *User type*(*Frequent user or Casual user*)

λ_{ijk}: *mean parameter in the negative biomial distribution*

θ: *over dispersion parameter in the negative binomial distribution*

X_{ijk}: *Explanatory variables*

Γ: *Gamma distribution function*

$\propto$: *Dispersion parameter in the dispersion test*

Result

Statistical Description

Out of 15,265,820 ETL trips that occurred during tolling times in 2016, 1,882,373 (12%) white double line violations were identified. It is expected that the actual number of violations is even greater. The number of violations by violation pattern is exhibited in Table 4, which shows that violations were concentrated on the southbound segments. The main reason for this is higher levels of traffic on the southbound segments compared with the northbound study segments of I-405.

When it comes to the relationship between violation frequencies and the characteristics of the road segments where the violations occurred, interesting trends are identified (Table 4). The segments with two ETLs tend to have a greater frequency of violations than the segments with one lane. One possible reason for this would be that the segments with two ETLs tend to have more weaving maneuvers from on-ramps or off-ramps, which may influence violation behaviors. In addition, there were fewer

Table 4. Statistics about Road Segment Characteristics

Violations	Violation category	# of ETLs	Segment length (mile)	Local access	Trips with violations	Percent
V1	Entry	2	0.75	1	28,868	1%
V2	Entry	2	2.38	0	137,675	7%
V3	Entry	1	1.14	0	127,350	6%
V4	Exit	1	1.14	0	175,786	9%
V5	Exit	1	3.41	0	323,575	16%
V6	Exit	2	0.94	0	386,822	20%
V7	Entry	2	0.38	1	42,207	2%
V8	Exit	2	1.82	0	364,225	18%
V9	Entry	2	1.82	0	321,790	16%
V10	Entry	2	1.82	1	63,479	3%

Note: The violation number corresponds to the number in the Figure 1. Local access 0 = without access; local access 1 = with access.

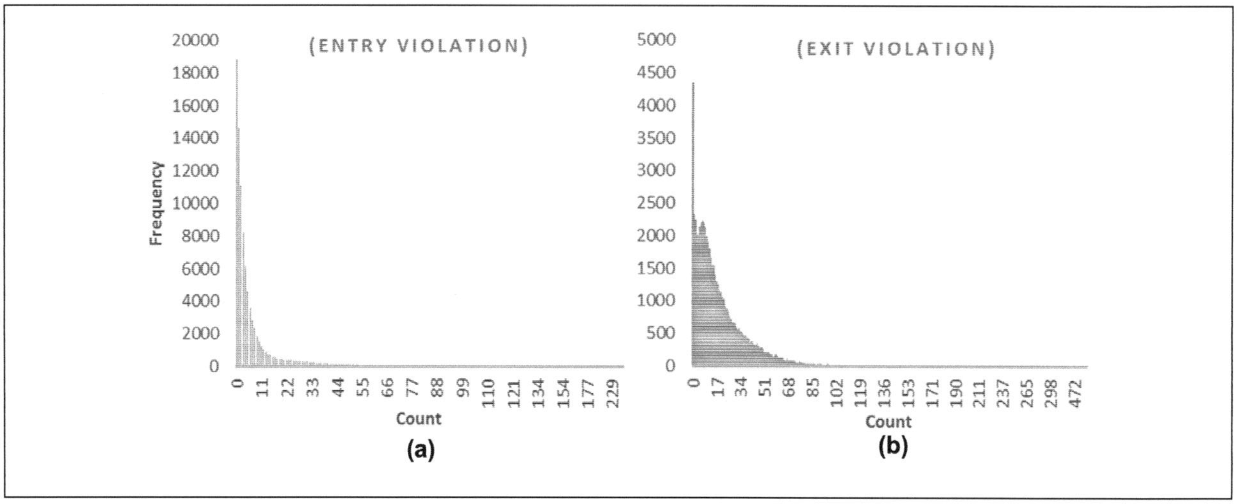

Figure 2. (*a*) Distribution of entry violation frequency; (*b*) distribution of exit violation frequency.

entry violations on segments with local access, which may imply violations are driven more by traffic conditions rather than local access. Another question of interest is whether violations increase as the road length between plazas increases. Although a relationship between road segment length and violation frequency was not apparent in the initial descriptive analysis, it is something that will be further examined in the model analysis.

One of the hypotheses of this study is that violation behavior on ETLs is significantly different when entering and exiting ETLs. In this dataset, there were 1,250,123 exit violations (63%) and 721,314 entry violations (37%). Figure 2 shows the difference in data distribution between entry and exit violations. Entry violation frequency has a standard deviation of 14.1 compared with 24.8 for exit violations. This implies that exit violations occurred in an extensive manner over time, compared with entry violation into ETLs. Based on this finding, it is plausible that traffic conditions or road design factors could influence violation behavior differently depending on whether the violation occurred when entering or exiting ETLs. Ideally, these violations should be modeled individually rather than modeling with pooled data.

This research is also noteworthy in that illegal behavior in ETLs was analyzed for frequent and infrequent users, as suggested by previous research (*9*, *10*). To do this, drivers were divided into two categories: frequent users and casual users. Vehicles identified by transponder (HOV and AVI types) were categorized as frequent users, while vehicles detected by license plate (IMG type) were defined as casual users. This study assumes that frequent users are more likely to obtain a transponder to take advantage of its convenience and cost savings. Those without valid transponders are more likely to be casual users who may be unfamiliar with the ETL system, such as rental car drivers or vehicles from out of town.

The number of violations committed by casual users was 1,447,123 or 28% of total trips conducted by casual users. There were 524,654 violations by frequent users, which was 5% of total trips conducted by frequent users. Figure 3 also shows that the mean of violation frequency for casual users is greater than the mean for frequent users, indicating that casual users commit access violations more frequently than frequent users. This may be because casual users include user segments who are unfamiliar with ETL access rules and may cross the white double lines without intending to violate.

This paper also classifies traffic status based on the speed of the GP lanes and the ETLs, using WSDOT's "state highway speed thresholds for congestion measurement" (*11*) (Table 5). Some 6% of observed speeds on ETLs were below 51 mph, indicating that the ETLs tend to provide optimal flow speed during tolling times and time reliability is high. The "severe congestion" status on the ETLs is a rare event and occurs when whole road segments, including the GP lanes, are congested. The traffic status variable was developed to see how traffic conditions on GP lanes or ETLs influence violation behavior.

Model Estimate Results

The statistical description clarified the different distributions for entry and exit violations, and the dispersion test suggested the NB model over the Poisson model for both datasets. Therefore, two NB models for entry violation and exit violation are estimated. The same set of traffic condition and user type variables is included in both models to allow for easy comparison. The variables in the road segment characteristics were chosen based on the result from the statistical descriptions for each model. The estimate results show the intuitive signs and values on variables across models.

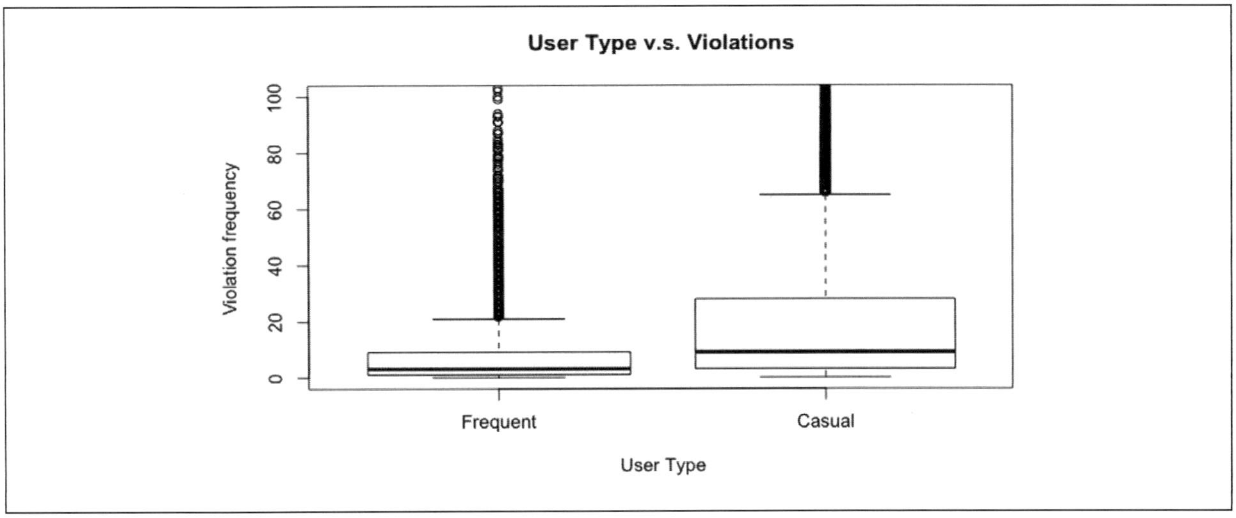

Figure 3. Distribution of access violation frequencies by the user type.

Table 5. Traffic Status Definition Based on the Speed

Speed criteria	Number of observations	Percentage
GP speed		
< 51 mph "state highway system delayed"	66,368	43
< 42 mph "state highway system congested"	24,740	16
< 36 mph "severe congestion"	16,308	11
ETLs speed		
< 51 mph "state highway system delayed"	9,748	6
< 42 mph "state highway system congested"	3,706	2
< 36 mph "severe congestion"	2,164	1

Traffic condition exhibits different effects across models. "Severe congested" traffic on GP lanes has a positive effect on entry violations but does not show a significant effect on exit violations. "System congested" ETL status negatively influences entry violations but has a positive effect on exit violations. This suggests that entry violations tend to occur to avoid congestion on GP lanes and exit violations occur to avoid congestion on ETLs, which is intuitive. It was statistically verified that entry and exit behavior differ depending on traffic conditions and that access violation behavior should be analyzed separately for entry and exit.

The effects of road segment characteristics and user type are also examined in the model. All variables included in the model were statistically significant and the estimated sign on all variables matched the trends seen in the statistical description. The model verified that access violations were more likely to occur on two-lane ETL segments and that entry violations were less frequent on local access segments. In addition, casual users were more likely to commit violations in both models.

The estimated results of the entry and exit violation models are shown in Table 6 and Table 7, respectively.

Summary and Conclusion

This research targeted access violation behavior crossing the white double lines which separate the ETLs from the GP lanes on I-405. The 10 violation patterns were identified from 2016 toll transaction data and the identified trips with violations were converted into 30-minute count data. The impact of traffic condition, user type, and road segment characteristics on access violation frequency was examined. The findings from the statistical description were introduced as explanatory variables in the model analysis. The data distribution of violation frequency shows the different patterns between entry and exit violations, and the dispersion test shows that violation frequency was over-dispersed in both entry and exit dataset. For this reason, two NB models for entry violation and exit violation were developed separately.

One of the major findings in the analysis is that entry and exit violation behavior differ depending on traffic conditions. The entry violation model found a positive effect for congested traffic on GP lanes and a negative effect for congested traffic on ETLs. On the other hand, congested traffic on GP lanes does not

Table 6. Estimate Result of NB Model for Entry Violation Frequency

Variables	Estimate	Std. error	z value	Pr (> |z|)	Significance
Intercept	0.36	0.01	37.67	< 2e-16	$p < 0.001$
User type:					
Dummy (0: frequent users, 1: casual users)	1.13	0.01	171.65	< 2e-16	$p < 0.001$
Traffic condition:					
GP traffic flow rate (veh/ln)[a]	0.46	0.004	106.07	< 2e-16	$p < 0.001$
GP speed dummy; "system delayed" (0: > = 51mph, 1: < 51mph)	0.73	0.01	93.74	< 2e-16	$p < 0.001$
GP speed dummy; "severe congested" (0: > = 36 mph, 1: < 36mph)	1.14	0.01	96.40	< 2e-16	$p < 0.001$
ETLs speed dummy; "system delayed" (0: > = 51mph, 1: < 51mph)	0.27	0.02	16.02	< 2e-16	$p < 0.001$
ETLs speed dummy; "system congested" (0: > = 42mph, 1: < 42mph)	−0.36	0.03	−13.36	< 2e-16	$p < 0.001$
Segment characteristics:					
The number of ETLs dummy (0: 1-lane, 1: 2-lane)	0.79	0.01	78.09	< 2e-16	$p < 0.001$
Local access dummy (0: else, 1: segment with access road)	−1.32	0.01	−181.96	< 2e-16	$p < 0.001$
Theta: 1.503 (std. err: 0.01)					
AIC (0): 554533					
AIC (β): 472120					

[a]Standardized value.

Table 7. Estimate Result of NB Model for Exit Violation Frequency

Variables	Estimate	Std. Error	z value	Pr (> |z|)	Significance
Intercept	1.16	0.01	131.09	< 2e-16	$p < 0.001$
User type:					
Dummy (0: frequent users, 1: casual users)	0.91	0.01	133.24	< 2e-16	$p < 0.001$
Traffic condition:					
GP traffic flow rate (veh/ln)[a]	0.41	0.004	98.26	< 2e-16	$p < 0.001$
GP speed dummy; "system delayed" (0: > = 51mph, 1: < 51mph)	0.42	0.01	51.75	< 2e-16	$p < 0.001$
GP speed dummy; "severe congested" (0: > = 36 mph, 1: < 36mph)	−0.02	0.02	−1.40	0.16	
ETLs speed dummy; "system delayed" (0: > = 51mph, 1: < 51mph)	0.62	0.02	34.09	< 2e-16	$p < 0.001$
ETLs speed dummy; "system congested" (0: > = 42mph, 1: < 42mph)	0.21	0.02	8.28	< 2e-16	$p < 0.001$
Segment characteristics:					
The number of ETLs dummy (0:1-lane, 1: 2-lane)	1.18	0.01	139.82	< 2e-16	$p < 0.001$
The segment length (mi)[a]	0.17	0.002	80.73	< 2e-16	$p < 0.001$
Theta: 1.588 (std. err: 0.01)					
AIC (0): 498552					
AIC (β): 464451					

[a]Standardized value.

have a significant effect on exit violations but congested traffic on ETLs increase exit violations. These results show that the drivers tend to commit entry violations to avoid congestion on GP lanes and commit exit violations when the ETL traffic gets congested. That is today, the characteristics of entry and exit violation are different in the context of traffic condition, which suggest that the entry violations and exit violations should be analyzed separately.

Another noteworthy finding is that the violation frequency is different between casual users and frequent users. The results show that casual users tend to commit violations more frequently over the whole tolling time period than frequent users. It is assumed that a large proportion of these violations are from unfamiliar users who cross the white double lines without intending to violate the rules. In order to reduce the number of unintentional violations, it would be beneficial to keep drivers informed of appropriate ETLs use using advertisements or clearer signage on the road.

The analysis provided new knowledge on the effect of road segment characteristics on access violations. The most significant factor was the number of ETLs; two-lane ETLs produced higher rates of entry and exit violations than one-lane segments. A potential reason for this could be weaving maneuvers from on-ramps or to off-ramps, which should be scrutinized in the future. In addition, violations were less likely to occur for ETLs with local road access. It is assumed that violation frequency depends on real-time traffic conditions rather than access to local roads.

In Washington State, enforcement for access violations is conducted through visual observation by WSP troopers. The findings can be utilized to improve enforcement planning to reduce access violation rates. For example, it may be more effective for enforcement officers to situate a patrol vehicle on the shoulder of a mainstream two-lane segment during periods of high congestion with the greatest possibility of access violations.

This analysis uniquely identified ETL violation patterns and developed new knowledge and findings on violation behavior. However, there are some limitations to this research. One example is that violation patterns were not comprehensive. This research focused on violations identified from the data collection characteristics and ignored other violations that would have occurred in the ETLs segments. In addition, violators that avoided paying toll fare entirely were not identified. In the future, the relation of toll evasion violations to toll rate will need to be discussed. Moreover, traffic conditions were only characterized by flow rate and travel speed in this research. Further analysis using traffic density, and traffic condition of on-ramp and off-ramp will be valuable to better understand the mechanism of ETLs violations.

This research introduced location-related factors as explanatory variables in the models and provided practical recommendations for WSP to effectively enforce access violations on ETLs of I-405. However, further analysis by separating models based on the spatial location will be required to examine the segment or road design impacts in detail, and to be able to transfer the analysis result to the operation or enforcement planning on other toll lane facilities.

Acknowledgments

This research would not be possible without toll transaction data from the WSDOT Toll Division and support from Eric Knigge. The author also acknowledges Dr. Yinhai Wang for primary advice and the UW STAR Lab members for their support in this research.

Author Contributions

The authors confirm contribution to the paper as follows: study conception and design: Hiroyuki Hinohara; data collection: Hiroyuki Hinohara, Eric Knigge (WSDOT); analysis and interpretation of results: Hiroyuki Hinohara, Yinhai Wang; draft manuscript preparation: Hiroyuki Hinohara. All authors reviewed the results and approved the final version of the manuscript.

References

1. Collier, T., and G. Goodin. *Managed Lanes: A Cross-Cutting Study*. FHWA-HOP-05-037. Federal Highway Administration, McLean, VA, 2004.
2. Washington State Department of Transportation. *I-5-SR16 Tacoma/ Pierce County HOV Program*. www.wsdot.wa.gov/Projects/PierceCountyHOV/. Accessed July 13, 2017.
3. Bastiste, J. R. *Washington State Patrol*. www.wsp.wa.gov/information/releases/2015_archive/mr092915.htm. Accessed July 13, 2017.
4. PRR, Inc. *Survey of Good To Go! Customer Users of I-405 Express Toll Lanes*. Survey Report. 2016. https://assets.documentcloud.org/documents/3113526/ETL-User-Survey-Report-Final.pdf.
5. Puget Sound Regional Council. *Puget Sound Regional Council. Transportation 2040 - Executive Summary*, 2010. https://www.psrc.org/sites/default/files/4web_finalt2040es.pdf.
6. Chou, C. S., E. Miller-Hooks, and X. Chen. Violations Modeling in Studies of Concurrent-Flow Lanes. *Transportation Research Record: Journal of the Transportation Research Board*, 2010. 2178: 147–155.
7. Jonathon, C., and M. Hallenbeck. *Examination of SR 167 HOT Lane Violation Patterns*. Research Report, WA-RD 766.1. Washington State Transportation Center (TRAC), 2011. http://depts.washington.edu/trac/bulkdisk/pdf/766.1.pdf
8. Washington State Department of Transportation. *I-405 Express Toll Lanes Entry and Exit Points*. www.wsdot.wa.gov/tolling/405/map/. Accessed July 13, 2017.
9. Adnan, S., A. Misra, and R. Guensler. High-Occupancy Toll Lane Decision Making Income Effects on I-85 Express Lanes. *Transportation Research Record: Journal of the Transportation Research Board*, 2015. 2531: 45–53.
10. Noah, G., and L. S. Brian. What Drives Decisions of Single-Occupant Travelers in High-Occupancy Vehicle Lanes? Investigation Using Archived Traffic and Tolling Data from MnPASS Express Lanes. *Transportation Research Record: Journal of the Transportation Research Board*, 2015. 2531: 45–53.
11. Lynn, P. *WSDOT's Handbook for Corridor Capacity Evaluation*, 1st ed. Washington State Department of Transportation. 2014, pp. 5. http://wsdot.wa.gov/publications/fulltext/graynotebook/CCR14_methodology.pdf.

The Standing Committee on Managed Lanes (AHB35) peer-reviewed this paper (18-03495).

Article

TRR
JOURNAL OF THE TRANSPORTATION RESEARCH BOARD

Technological and Human Factor Considerations for Priced Managed Lane Traveler Information Systems

Transportation Research Record
2018, Vol. 2672(14) 10–19

DOI: 10.1177/0361198118758680
journals.sagepub.com/home/trr
SAGE

Nicholas S. Wood[1], Chris Simek[1], Susan T. Chrysler[2], Jeff Kaufman[3], Shawn Turner[2], and David Fink[4]

Abstract

Travelers have many unique informational requirements to be able to navigate priced managed lanes. These demands often relate to specific managed lane features (e.g., access points, toll price) in addition to information about major traffic incidents and lane closures. Accommodating these needs with traditional roadway signage is a particular challenge given concerns about overloading and distracting drivers. This paper summarizes an investigation into traveler information systems for managed lanes, by considering a national review of current agency practices, a traveler survey, and an assessment for integrating advancing technologies. The national review found that many agencies vary considerably in pricing structure, number, and placement of priced destination points, and online availability of real-time toll information. A travel survey of 866 Texas-based respondents indicated that drivers prioritize information about traffic incidents and lane closures over toll price data (94% and 88% versus 41%, respectively). A higher share of respondents wanted to see travel time and incident alerts on in-vehicle devices, compared with a more statistically significant desire for destination and toll rate information on roadway signs. Most respondents use smartphone applications and mapping websites for pre-trip planning purposes (79% and 65%, respectively) compared with TV and radio reports (13%). Comparatively, prior research published 5 years earlier found that radio was a highly influential media in influencing behavior. This paper suggests a pathway for agencies to adopt a flexible approach for sharing essential data with third-party entities, based on the general transit-feed specification used for transit.

Managed lanes are inherently complex facilities that require advanced traveler information systems to serve the users of those lanes. These intricacies fundamentally stem from the diverse variety of managed lanes with critical operational and geometric differences. Each managed lane has a definition distinguished by a set of strategies that commonly include restrictions for select user groups (e.g., carpools, low-emission vehicles) and provisions for intermediate access (e.g., continuous, limited access). Other strategies entail the integration of tolling, transit services, select hours of operation, and direct access ramp location. Tolling is a frequently used strategy that project sponsors use to refine performance, either through static or dynamic tolling. Common managed lane types include high-occupancy vehicle lanes, high-occupancy toll (HOT) lanes, and express toll lanes.

The varied types of managed lanes demand a communication system to support a wide diversity of traveler information needs. Project sponsors have to post information that addresses the needs of all potential users. Traveler needs include the number of required carpool occupants, tolls, compatible toll transponders, destinations, travel times, crash indication, and open/closed status. Currently, operators rely mostly on signage to convey related information, but signage cannot practically display all of the necessary information to navigate a managed lane. Additionally, the principles of positive guidance dictate that not all information needs to be present at all locations (*1*).

The overall complexity of managed lanes has increased given the introduction of managed lane networks. Historically, managed lanes existed as a concept for single and distinct corridors, providing preferential access for users who travel from one endpoint to another. These initial corridor-based systems were successful, leading to the planning and construction of networks of managed lanes. A regional managed lane network can either exist (1) by direct interconnections through complex interchanges, or (2) distinct systems on separate corridors marked by common trailblazers and signage. However, different types of managed lanes can exist within a network due to various project timelines, sponsors, goals, and other factors.

[1]Texas A&M Transportation Institute, Austin, TX
[2]Texas A&M Transportation Institute, College Station, TX
[3]Texas A&M Transportation Institute, Houston, TX
[4]Houston-Galveston Area Council, Houston, TX

Corresponding Author:
Address correspondence to Nicholas S. Wood: nickwood@tamu.edu

Travelers who use multiple corridors have to understand those differences for efficient travel management. Additionally, existing signage and different communication methods will become more prone to ineffectiveness as network complexity increases. New technology, including advancements in traveler information and connected vehicles, may provide a solution to the problem.

National Review

A state-of-the-practice review investigated the overall factors that influence traveler perceptions of information systems, in addition to an analysis of existing and proposed managed lane networks within the United States (U.S.). The review used three different techniques for gathering current practices: (1) reviewing existing literature, (2) in-person site visits, and (3) stakeholder interviews. Overall, critical issues were assessed and summarized for specific cross-cutting topics related to signing, operating policies, understanding traveler behavior, and geometric design. Seven detailed case studies examined how traveler information systems were applied and working within a regional context. Each case study detailed the roles and responsibilities of key stakeholders, the type of pricing system, and type of pre-trip, en-route, and post-trip information providers to travelers of the managed lane network.

In-Field Physical Characteristics

At first glance, geometric design can have a significant influence on travelers' perceptions of managed lanes. Roadway markings designate access locations, and signage summarizes essential operating rules. Critical geometric attributes that influence traveler information systems for managed lanes include (*2–5*):

- Consistency. A design that provides a sense of continuity allows more familiarity to make decisions with less hesitation and uncertainty.
- Roadway signage. The type, arrangement, and frequency of signage can inform, encourage, or discourage usage.
- Ramps. The type and spacing of ramp orientation can influence the ability of drivers to decide at access points.
- Separation. The separation (e.g., barrier, pylon, striped buffer) between the managed and the general-purpose lane influences traveler perceptions of access. If the managed lane appears to be uninviting or inaccessible, users will become discouraged from using that facility.
- Access. Frequency and type of access to a managed lane influences traveler behavior according to whether the managed lane provides for intermediate, continuous, or direct access to park-and-ride facilities.
- Tolling equipment. The presence of tolling equipment influences behaviors related to free access (e.g., free and discounted carpools), user acceptance, and economic incentives or disincentives.

Behavioral Characteristics

A plethora of existing research assessed and summarized how travelers react to available user information systems for managed lanes. Young and old travelers have different tendencies for obtaining information, with younger users being receptive to more technically advanced applications (*6,7*). The most prevalent medium for obtaining traveler information came from radio reports, as cited by existing research (*8–10*). Some reports also stated radio was the most effective medium for inciting travel behavior changes (*11*). Many travelers frequently use interactive maps to obtain information (*12*). Reliability and travel time savings were the two most commonly cited reasons why travelers choose to travel in a managed lane, in addition to gaining a feeling of greater safety and comfort by selecting a longer, more predictable route (*11,13,14*). Overall, travelers overestimate travel time savings associated with managed lane use (*15,16*). While the connection between mental variables and choices made in a managed lane context was weaker than expected, some psychological questions are useful in predicting managed lane behavior.

Existing and Planned Regional Systems

A primary focus of the state-of-the-practice review focused on an assessment of seven regional managed lane networks, which consisted of Dallas/Fort Worth, Houston, Minneapolis-St. Paul, South Florida (Miami), Seattle, San Francisco (Bay Area), and Washington, D.C./Northern Virginia. These systems are not comprehensive but represent a national composite that differs in relation to regional goals, project sponsors, and physical scale. The review consisted of stakeholder interviews with project sponsors, regional intelligent-transportation-system operators, and mid-level state department of transportation leadership. Additional information that augmented the analysis came from an online examination of publically available traveler information sources, including web pages and pamphlets. Table 1 summarizes the seven networks by types of pricing structure, ability to view toll rates online, and the number of priced destinations on overhead signage. Those three elements comprise the more distinguishing characteristics regarding travel information for regional managed lane systems.

Pricing Structure. The complexity of traveler information derives from, in part, the wide variety of pricing structures in practice throughout the U.S. A pricing structure is the set of definitions that describe the toll charges to travel a set distance, or between a series of points. Managed lanes can either have a single price for entry, a per-mile charge (users charged

Table 1. Key Characteristics for Regional Priced Managed Lanes

Network	Current number of priced corridors and total miles	Pricing structure	Can view toll rates online?	Number of priced destinations on toll rate sign (or single sign bridge)
Dallas/Fort Worth	6 corridors/79 miles	Destination-based segment pricing	No	1 or 2 (depending on facility)
Houston	6 corridors/97 miles	Single entry price	Yes (fixed, time-of-day rates)	1 (METRO HOT), 3 (Katy MLs)
Minneapolis-St. Paul, Minnesota	3 corridors/31 miles	Destination-based segment pricing	No	2
South Florida	2 corridors/17 miles	Destination-based segment pricing	No	3
Seattle, Washington	2 corridors/26 miles	Destination-based segment pricing	Yes	3
San Francisco Bay Area	4 corridors/72 miles	Zone-based pricing (depending on facility)	Some facilities	2
Washington, D.C./ Northern VA	2 corridors/43 miles	Destination-based segment pricing	Yes	3

a rate of US$/mile), or a charge based on traveling to specific destinations or zones. Operators use different pricing structures because of established regional policies, geometric design limitations, demand for specific origin–destination pairs, installation of tolling equipment, or enforcement capability. The variety of pricing structures enables flexibility for operators, but can lead to complexity for the end user. All the facilities examined, except for Houston, use dynamic pricing that varies in real-time based on demand. The toll rates for the I-10 Katy Freeway Managed Lanes, and the METRO HOT Lanes come from an existing, fixed time-of-day schedule (*17,18*).

From the review, most managed lanes operate with a pricing structure that varies based on the specific destination selected by the traveler, or destination-based segment pricing. The unit or per-mile toll price is not the same for all tolled travelers. For example, a traveler with destination pricing could pay US$9.00 to travel 9 mi, but another driver who goes 17 mi could be charged US$12.00— due to a lower per-mile charge to drive a longer distance. Facilities that apply a uniform rate apply link-based pricing, where the toll rate adjusts based on the location with the highest demand, which is applied evenly across the entire corridor. The managed lanes in the Washington, D.C. area have destination-based pricing, in addition to facilities in the Atlanta, Georgia, region (*19*), (Virginia Department of Transportation, personal communication, Jul. 1, 2016).

Presently under construction, the I-880 Express Lanes in Alameda County, California, will have one of the most complex pricing structures in operation. Roadway signs will generally display toll rates to the nearest exit point in addition to a spot farther downstream that may not be the second-closest exit point. However, some zones will have continuous access with no designated exit point. The Metropolitan Transportation Commission, responsible for managing the I-880 Express Lanes, justifies the zone-based pricing structure by referring to the high levels of expected demand for some Express Lane segments (Metropolitan Transportation Commission, personal communication, Jul. 8, 2016). In contrast, the more fundamental METRO HOT Lanes in Houston only charge a single price for an entering that does not vary based on distance traveled (*18*).

Number of Priced Destinations on Signs. The number of priced destinations on overhead signage tended to vary by specific networks. Managed lane systems in Seattle and South Florida are currently planning to display prices for three distinct destination points (Washington State Department of Transportation, personal communication, Jun. 30, 2016; and Florida Department of Transportation, personal communication, Jul. 1, 2016). During the stakeholder interview, the Washington State Department of Transportation reported that users often confused the three prices as being additive, meaning, travelers thought they paid all three prices instead of one, if they traveled the long distance. In comparison, the I-35W North Tarrant Express in Fort Worth will only display one price on overhead signage to the next exit. The rationale offered for signing prices on the North Tarrant Express was the belief of the private concessionaire that displaying higher prices for longer distances would discourage use of the managed lanes (Texas Department of Transportation, personal communication, Apr. 11, 2016).

Online Toll Rate Accessibility. Another practice that varied by facility was the capability of the end user to view real-time toll rates online or through a mobile application. Managed lane facilities in the San Francisco, Seattle, and Washington, D.C. regions allow users to visit a website to view real-time tolling information (*20–22*). However, operators in the Minneapolis-St. Paul and South Florida regions do not show real-time tolls due to concerns about traveler complaints if rates differ from the online information (Minnesota Department of Transportation, personal communication, Jul. 8, 2016; and Florida Department of Transportation, personal communication, Jul. 1, 2016). Additionally, operators in San Francisco and Seattle permitted outside parties to access live data-feeds that provide real-time tolling rates, either by obtaining code on a third-party website (e.g., GitHub) or through enhancements to an existing 511 system (*23,24*). Private entities had to sign a data-sharing agreement with the operating agency to have access.

Travel Survey

A human factors assessment considered how travelers value and understand informational needs for navigating complex managed lanes and managed lane networks. The research used two different techniques for investigating traveler information needs and perceptions: focus groups and online surveys. The focus groups cast a broad net to gather findings through open-ended, small group discussions held in the Dallas/Fort Worth, Houston, Austin, and College Station regions. College Station served as the control group as a location with no operating managed lane facilities. Thirty-two individuals participated in one of the four focus groups. Based on the focus group findings, researchers identified the following as fundamental research questions that the online survey should address:

- What communication mechanisms do people use to plan a route?
- What features of trip planning systems do drivers prefer?
- What is the preferred communication mechanism for pre-trip and en-route information?
- What specific type of regulatory and guidance information would drivers like to see on road signs compared with in-vehicle devices?

Survey Development and Deployment

The goal of the survey was to identify what information travelers use and desire for planning managed lane trips. The survey relied on a web-based instrument to solicit responses from respondents residing in the Dallas Fort/Worth, Houston, and Austin regions. Responses from travelers who did not live in one of those three regions were set aside for further analysis. The self-administered questionnaire contained approximately thirty questions, customized for different respondents based on the location of residence and use of technology. Sections of the survey asked where travelers made a majority of their trips, searched for travel information, and reasons for using managed lanes.

The customized survey instrument allowed respondents to answer questions that best fit their level of understanding. Tailored surveys helped to reduce the time to take the survey, which helped increase the overall response rate. Answers from the first set of questions guided the selection and order of subsequent questions. For example, respondents did not see questions about managed lanes in Austin if they indicated that most trips occurred in Houston. Another question asked respondents if they traveled in a managed lane. If respondents did not travel in a managed lane, a follow-up question asked them to select a reason for this from a set of pre-established answers. Furthermore, these respondents did not see subsequent questions about technology. Therefore, the use of customized surveys led to more responses for certain questions than others, resulting in different sample sizes for individual questions.

Survey recruitment occurred through various media, including the use of traditional media sources and newer social media services. Traditional recruitment methods included distribution of press releases to newspapers and radio stations. Newspapers in Austin and Dallas/Fort Worth published online articles about the survey, and research staff conducted interviews with a radio personality in Dallas/Fort Worth. Social media promotion occurred primarily through Facebook and Twitter. Facebook indicated that 7,590 users of their system saw a promotion for the survey during the open response period. Twitter reported that 58,953 users saw the survey promotion, resulting in 109 clicks to the survey web address. Houston TranStar, the traffic management center in Houston, also advertised the survey on their main web page by using a small clickable green button that appeared in the upper right-hand corner.

A limitation of the methodology was the reliance on self-selecting respondents for an Internet survey, and the results may have undercounted travelers with limited to no understanding of managed lanes. Resources available for the survey effort prohibited a more robust and rigorous sample design. As such, researchers implemented nonprobability sampling techniques, a commonly used practice for developing a methodology that considers research budget constraints. A significant downside of this sampling method is the lack of reliable sampling errors.

Results

The open response period for the survey was February 2, 2017 to March 12, 2017. After examining the 1,128 responses, 190 surveys were removed due to partial completion. The remaining 938 completed surveys included responses from Houston (579 surveys), Dallas/Fort Worth (159 surveys), and Austin (128 surveys). The analysis only focused on these 866 surveys from the three indicated metro regions. The 72 remaining surveys came from respondents who resided outside the three selected metropolitan regions. The median time to complete the survey was approximately 7 min. This paper highlights results related to technology preference,

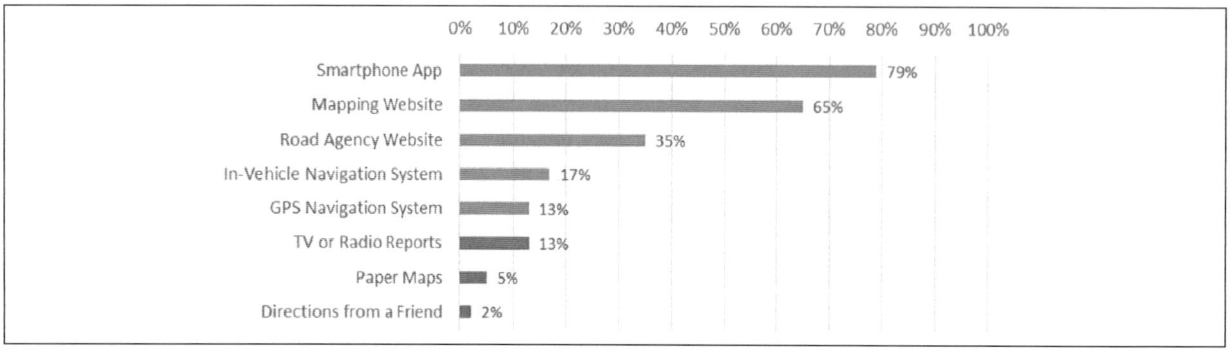

Figure 1. Use of traveler information often or very often (n = 866).

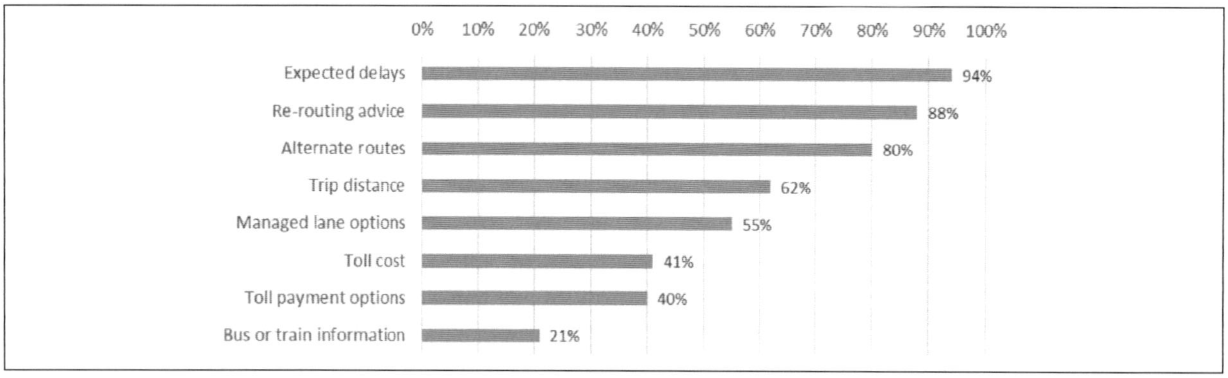

Figure 2. Travel-planning elements viewed as important or critical (n = 100).

information desired, reasons for not using managed lanes, and choices for in-vehicle or roadside signage of information.

Within the survey, respondents initially saw a list of traveler information sources (e.g., TV, radio reports, smartphone apps) and asked how frequently they used each in the process of selecting routes. Figure 1 presents the proportional distribution of sources used often or very often. These data suggest that smartphone apps and mapping websites are the most frequently used sources of traveler information. These results corroborate with similar findings from a 2016 traveler information survey conducted on behalf of the Texas Department of Transportation, Austin District (*25*). For that study, a majority of respondents (68%) indicated they prefer to use smartphone apps to obtain traveler information for commute trips. This proportion increased to 73% for noncommute work or school trips. The current study, as shown in Figure 1, found that 79% of respondents used smartphone apps.

For the following questions, respondents imagined they were planning a trip at a time in the near future (either that day or later in the week). They saw a list of various types of traveler information and were asked to indicate how important each type of information was as a travel-planning tool. Figure 2 presents the proportional distribution of sources believed to be critical or essential. The highest proportion of all respondents felt that expected delays and re-routing advice were critical or important (94% and 88% of respondents, respectively). Conversely, the data element with the lowest share of critical or important responses was information about buses, trains, and alternative modes of transport (23% of respondents).

The survey also provides input into reasons why drivers are not using managed lanes. Respondents who reported never using a managed lane were presented a question to identify the most important reason for not using those facilities. Figure 3 shows that managed lane nonusers choose not to use the facility because the lane does not go to their destination. Nearly a third (31%) of nonusers chose destination location as a response, while almost one-quarter (23%) indicated they are opposed to toll lanes in general. One in five respondents was wary of using the facility because they were not sure about where to exit.

Respondents who make managed lane trips saw a question about the most important factor in their decision to use those facilities. More than a third of all managed lane users (36%) stated that "traffic conditions that I observe in the regular lanes" was the most important factor. The second highest indicted factor, by slightly less than one-quarter of managed lane users (23%), was "traffic reports on a smartphone or in-vehicle GPS." Respondents expressed that radio

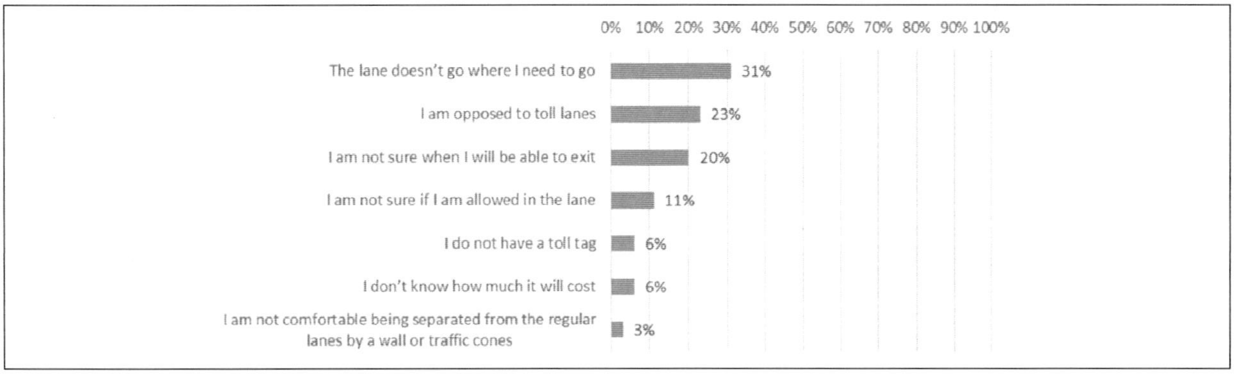

Figure 3. Reasons managed lanes are not used (n = 224).

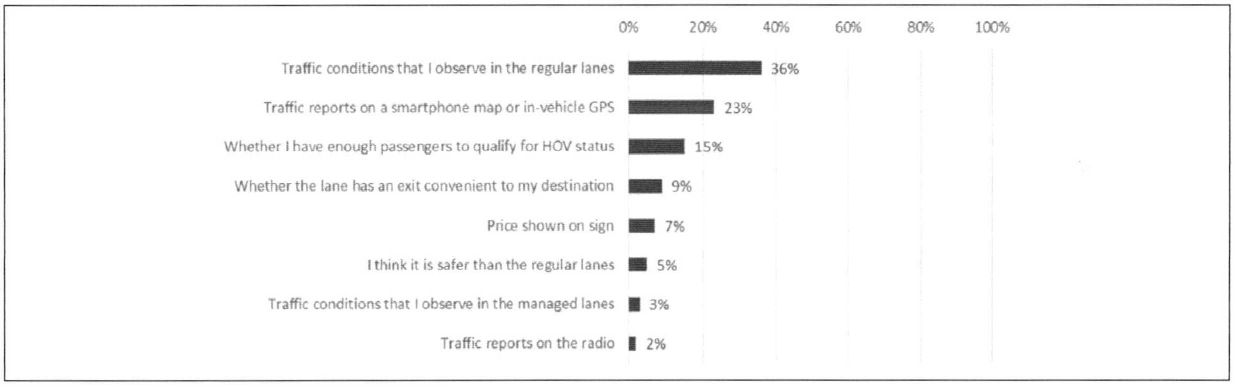

Figure 4. Single most important decision factor for managed lane use (n = 635).

traffic reports were the least important factor (2%). Figure 4 shows the proportional distribution of respondents who indicated a specific factor as being the most important in their decision to use the managed lanes.

A further chi-square analysis suggested a statistically significant relationship between the single most important factor in deciding to use a manage lane and geography. Over half (55%) of Austin managed lane users reported that traffic conditions observed in the regular lanes was the most important factor, relative to about a third (32%) of Houston managed lane users that reported the same (Pearson's chi-square value = 8.678, $p < .01$). Austin respondents were likely to have less experience with managed lanes, given the first type of that facility opened for operation only 4 months before the survey began. Managed lanes in Houston, in contrast, have operated for almost four decades.

One question asked respondents to imagine a near-term scenario in which traveler information would show on their smartphone, a navigation system, or an in-vehicle device. Transmitting more data to electronic devices might lessen the need for road signs to be installed. For this question, respondents saw several elements of traveler information and were asked to indicate whether they thought they should appear on a roadway sign, or an in-vehicle device. Figure 5 presents a summary of responses to this question. "Entrance location to the managed lane," and "number of occupants required" were the two elements mentioned by the greatest proportion of respondents as being best suited for display on road signs. Conversely, "alerts about crashes, construction, or other accidents" and "traffic condition status" were elements mentioned by the highest proportion of respondents as being best suited for display on in-vehicle devices.

The survey asked the 306 users of managed lanes about their first use of the facility. The majority (66%) reported that only road signs influenced their decision-making process, without any other information about the lanes. This finding highlights the importance of road signs for first-time, unfamiliar users. The results shown in Figure 5 illustrate that drivers desire static guidance and regulatory information on road signs, whereas information related to traffic conditions or information specific to their trip should be shown on in-vehicle displays.

Discussion

According to a 2017 report published by the Pew Research Center, approximately three out of every four Americans own a smartphone. This proportion had doubled since 2011 when the Pew Research Center first began tracking this data

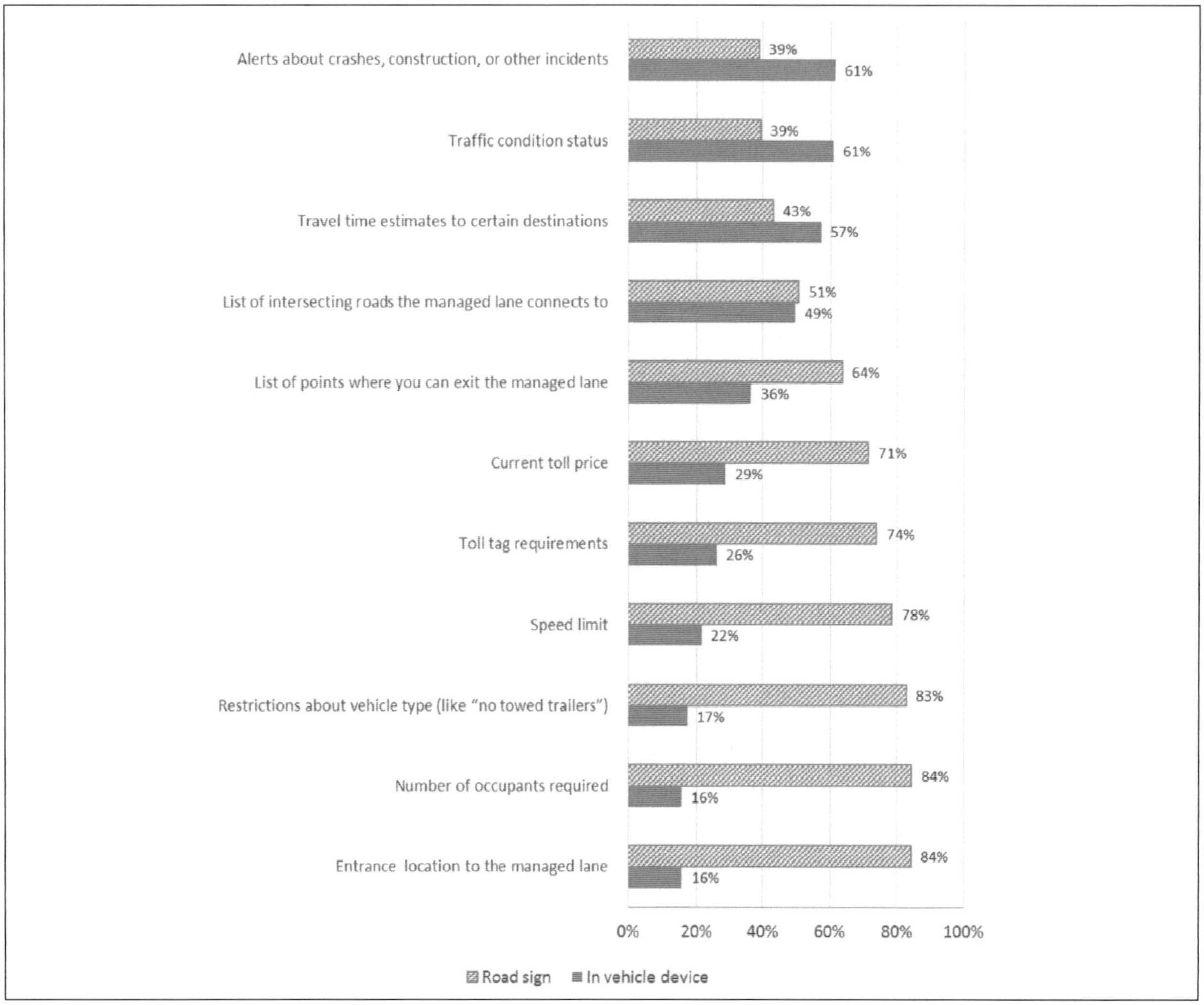

Figure 5. Where managed lane information should be displayed (n = 850).

point. The report cites the statistic that seven out of 10 Americans use social media and half of all Americans own a tablet computer (*26*). The Pew report corroborates the finding from the survey population that shows a strong preference for smartphone apps and mapping websites. Similarly, some technology companies have seen the demand for enhanced traveler information services by providing applications with real-time information that incorporates expected delays, re-routing advice, and alternate routes.

Existing driver knowledge and perceptions can have a significant influence on the efficacy of a travel information system. Some drivers do not travel in the managed lane at all due to an adverse opinion about tolling. Informational campaigns can change perceptions, mainly if the material presented focuses on showing how the benefits will come to fruition (revenue generation, decreased congestion in general-purpose lanes, etc.). However, changing public opinion is a challenging and resource-intensive process. Additionally, communicating the concept of managed lanes is a challenge that stems from the unique nature of those facilities.

A significant proportion of nonusers (20%) who do not travel on priced managed lanes, indicated they did not use the facility because they did not know where to exit the lane. As such, it is logical to assume that focusing outreach efforts on managed lane exit locations may be a worthwhile endeavor. A potentially effective way of doing so is to physically display exit information on facility signage, as two-thirds of managed lane users based their decision to do so strictly on information they obtained via road signage. Any provision of in-vehicle information will need to consider driver distraction and information overload issues.

Integrating Emerging Technologies

Transportation entities entrusted with managing traveler information systems have to understand the continually evolving ecosystem of products, services, and technologies. Prior research, as recent as 2013, have indicated that many travelers—almost half—listen to radio reports that are highly suggestive for influencing travel behavior (*8–10*). However, the 2017 travel survey as described in the preceding section found that only 2% of representative travelers feel that radio is the most important factor for influencing their use of a priced managed lane.

Knowing these facts, researchers suggested an implementation pathway for entities to augment and improve their existing systems by principally focusing on the sharing of data. A path that emphasizes the selection or preference of specific technologies will likely lead to reduced traveler perceptions and acceptance due to lagged deployment. Data flexibility allows their application across a range of technologies, including roadway signs, in-vehicle devices, and connected vehicles. Researchers conducted stakeholder interviews with various private-sector data companies to understand the factors that influence their decision to incorporate managed lane traveler information.

Private-Sector Data

Private-sector data providers are entities that produce real-time traffic information using data aggregated from various sources. These sources encompass location-based services that include smartphones, GPS devices, Bluetooth sensors, and other connected devices. Often, these entities do not produce hardware or create maps, but instead, provide data for use by transportation agencies or other private companies. Data from private entities commonly include elements related to speed, congestion, traffic incidents, and roadway geometry. Additionally, private-sector providers create application programming interfaces and develop specialized analytical services for customers, including hotspot analyses, parking tools, and weather indicators.

Researchers conducted a series of in-depth interviews with private-sector companies that develop maps and provide traveler information services. The purpose of the discussions was to identify current practice and investigate the potential for enhancing their products with shared, open data from transportation agencies. Many of the private companies interviewed do not emphasize or focus solely on providing information for managed lanes. Some companies expressed a willingness to incorporate managed lane information into their products, whereas others felt that this type of material did not serve their primary customers, such as freight and long-distance deliveries. One major private-sector provider believed that dynamically priced facilities were not widespread or common in the U.S. From the interviews, researchers noted these key findings for consideration in developing a data-sharing platform for managed lanes.

- Limitations exist for defining lane-specific information. Companies currently have difficulty assessing speed data relative to small dimensions, primarily if speed relies on quality GPS data. Managed lanes that have barriers, or a wide separation, tend to appear in online maps more often than facilities narrowly separated by a striped or painted buffer. Some providers have started internal research efforts to gather data about the number of lanes, lane width, and locations with lane splitting. The companies also noted that research was part of a larger project to develop maps for future automated vehicle applications.
- Companies prioritize information related to routing decisions, lane closures, detours, and major traffic incidents. This finding from the private-sector interviews corroborates the results from the traveler survey.
- Concerns exist about the reliability and accuracy of data published by public-sector transportation agencies. Some private entities search the Internet for toll rates and other information, only to discover incomplete and dated web pages. Clear documentation should note which data elements are relatively static or change within a specified frequency.
- Definitions are not consistent across many different facilities nationally and within regions. Private companies have expressed confusion about how agencies charge tolls and apply unique operational rules (e.g., reversible direction, hours of operation, exclusive toll discounts). Overall, data from agencies should provide timely, accurate, and structured information.

Potential for Future Data Feed Specification

A possible solution might incorporate a data-transfer protocol similar to real-time General Transit Feed Specification (GTFS) services currently used for transit. Real-time GTFS allows transit providers to display the present locations of its buses, estimated times of arrival at transit stops, and service updates for the system. The transit provider receives the vehicle geo-coordinates, and the position transmits to the map as defined by the feed specification (*27*). Real-time GTFS relies on two different approaches for transmitting data known as fetch, and push. The fetch method consists of an agency hosting the GTFS real-time feed on its server (or use of a web application server), and the mapping service provider can pull the data every 30 s. The push approach entails the agency setting up automatic data uploads to outside entities (*28*).

Conclusions

This paper presents the results of an assessment of traveler information for priced managed lanes. The steps of the research consisted of assessing existing agency practices, seeking feedback from drivers, and considering improvements that incorporate advancing technologies. Overall, current practice tends to vary nationally and within individual regions, as influenced by how different agencies approach pricing structures, the number of priced destination points on overhead signs, and online availability of real-time prices. A significant difference between these systems is the assessment of toll charges: some operators assess fees based on passing single points versus others that consider origin–destination pairs.

Traveler surveys showed that managed lane users desire particular information for specific media, which can change over time. Most survey respondents do not value toll price information as highly as data about traffic incidents and lane closures. A high 94% and 88% of respondents indicated that information about expected delays and re-routing advice was important or critical, compared with 41% for toll price information. However, the primary needs and sources of traveler information change regularly. A higher share of travelers wanted to see travel time and incident alerts on in-vehicle devices, compared with a more statistically significant desire for destination and toll rate information on roadway signs. Most respondents use smartphone applications and mapping websites for pre-trip planning purposes (79% and 65%, respectively) compared with TV and radio reports (13%). Comparatively, prior research published 5 years earlier found that radio was a highly influential media in influencing behavior.

A limitation of the methodology was the reliance on self-selecting respondents for an Internet survey, and the results may have undercounted travelers with limited to no understanding of managed lanes. Resources available for the survey effort prohibited a more robust and methodologically rigorous sample design.

The authors suggest a pathway for agencies to adopt a flexible approach for sharing essential data with third-party entities, using the general transit-feed specification. This approach limits the operating agency to principally managing the outflow of critical data, without endorsing a specific medium that can quickly change in relevance. A key implementation challenge requires coordination between agencies to agree upon a common framework for providing both static and dynamic information. Organizations might consider the relative essential elements of priced managed lane data as a precursor for developing platforms to support automated and connected vehicle applications.

Acknowledgments

The Texas Department of Transportation supported this research as part of the project Communicating Information for Traveling on Managed Lane Networks. The authors thank the following project panel individuals: Darrin Jensen, Theresa Poer, Flor Tamez, and John Nevares from the Texas Department of Transportation; Dan Lamers, Natalie Bettger, and Amanda Wilson from the North Central Texas Council of Governments; and Nader Mirjamali and Hameed Merchant from Houston METRO. The authors also recognize the following individuals for assisting with individual tasks: Maarit Moran, Lisa Green, Marcus Brewer, Tina Geiselbrecht, Trey Baker, Michelle Canton, Mike Vickich, and Tony Voight.

References

1. Lunenfeld, H., and G. J. Alexander. *A User's Guide to Positive Guidance*, 3rd ed. Report FHWA-SA-90-017. FHWA, U.S. Department of Transportation, September 1990.
2. Brydia, R. E., and S. Song. *Interoperability Issues on Managed Lane Facilities*. Report FHWA/TX-05/0-4160-18. Texas A&M Transportation Institute, College Station, TX, 2004. http://tti.tamu.edu/documents/0-4160-18.pdf. Accessed July 1, 2017.
3. Sas, M., S. Carlson, E. Kim, and M. Quant. Consideration for High Occupancy Vehicle (HOV) to High Occupancy Toll (HOT) Lanes Study. Report FHWA-HOP-08-034. HOV Pooled-Fund Study, Federal Highway Administration, Washington, D.C., 2007. http://www.ops.fhwa.dot.gov/publications/fhwahop08034/fhwa_hot_lane.pdf. Accessed July 1, 2017.
4. Stanitsas, P., J. Hourdos, and S. Zitzow. *Evaluation of the Effect of MnPASS Lane Design on Mobility and Safety*. Report MN/RC 2014-23. Minnesota Traffic Observatory, Department of Civil Engineering, University of Minnesota, Minneapolis, MN, 2014.
5. Fitzpatrick, K., M. A. Brewer, and S. Venglar. 2003. *Managed Lane Ramp and Roadway Design Issues*. Report 0-4160-10. Texas A&M Transportation Institute, College Station, Texas, 2003.
6. Kristof, T., M. Lowry, and G. S. Rutherford. *Assessing the Benefits of Traveler and Transportation Information Systems*. Washington State Transportation Center (TRAC), 2005.
7. Peirce, S., S. Puckett, M. Petrella, P. Minnice, and R. Ray. Effects of Full-Facility Variable Tolling on Traveler Behavior: Evidence from a Panel Study of the SR-520 Corridor in Seattle. Volpe National Transportation Systems Center, 2014.
8. Tseng, Y-Y., J. Knockaert, and E. T. Verhoef. A Revealed-Preference Study of Behavioral Impacts of Real-Time Traffic Information. *Transportation Research Part C*, Vol. 30, 2013, pp. 196–209.
9. Petrella, M, P. Minnice, and J. Lappin. Traveler Use of and Response to Real-Time Traffic and Traveler Information-Evidence from Integrated Corridor Management Traveler Surveys in Dallas, Texas, and San Diego, California. *Transportation Research Record: Journal of the Transportation Research Board*, 2014. 2423: 44–51.
10. Petrella, M., S. Puckett, S. Peirce, P. Minnice, and J. Lappin. *Effects of Full-Facility Variable Tolling on Traveler Behavior: Evidence from a Panel Study of the I-85 Corridor in Atlanta*. Volpe National Transportation Systems Center, 2014.
11. Kuhn, B., L. Higgins, A. Nelson, M. Finley, G. Ullman, S. Chrysler, K. Wunderlich, V. Shah, and C. Dudek. *SHRP 2 Report S2-L14-RW-1: Effectiveness of Different Approaches to Disseminating Traveler Information on Travel Time Reliability*. Transportation Research Board of the National Academies, Washington, D.C., 2014.
12. Wang, X., A. J. Khattak, and Y. Fan. Role of Dynamic Information in Supporting Changes in Travel Behavior-Two Stage Process of Travel Decision. *Transportation Research Record: Journal of the Transportation Research Board*, 2009. 2138: 85–93.

13. Du, J., F. Guo, and H. A. Rakha. Evaluation of High Occupancy Toll Lane Usage by Single Occupancy Vehicles. Presented at 95th Annual Meeting of the Transportation Research Board, Washington, D.C., 2016.
14. Jin, X., M.S. Hossan, and H. Asgari. *Investigating the Value of Time and Value of Reliability for Managed Lanes*. Report BDV29-977-12. Florida Department of Transportation, Miami, FL, 2015.
15. Devarasetty, P. C., M. Burris, J. McDonald, and G. J. Munoz. Can Psychological Variables Help Predict the Use of Priced Managed Lanes? *Transportation Research Part F: Traffic Psychology and Behavior*, Vol. 22, 2014, pp. 25–38.
16.Carrion, C., and D. Levinson D. Value of Travel Time Reliability: A Review of Current Evidence. *Transportation Research Part A*, Vol. 46, 2012, pp. 720–741.
17. *Harris County Toll Road Authority*. Toll Rate Schedule – Katy Freeway Managed Lanes. https://www.hctra.org/Content/hctra3rdPartyPages/katymanagedlanes/media/Katy_Toll_Sched.pdf. Accessed July 1, 2017.
18. *HOV/HOT Lanes Frequently Asked Questions*. Metropolitan Transit Authority of Harris County, Texas. http://www.ridemetro.org/Pages/HOVandHOTLanesFAQs.aspx. Accessed July 1, 2017.
19. *State Road and Tollway Authority*. I-85 Toll Rate Pricing. http://peachpass.com/peach-pass-toll-facilities/i-85-toll-rate-pricing. Access July 1, 2017.
20. *Santa Clara Valley Transportation Authority*. How to Use the Silicon Valley Express Lanes. http://www.vta.org/getting-around/using-express-lanes. Accessed July 1, 2017.
21. *Washington State Department of Transportation*. I-405 Express Toll Lanes—Current Rates. http://www.wsdot.com/traffic/seattle/flowmaps/I-405Toll.aspx. Accessed July 1, 2017.
22. *Transurban (USA) Operations Inc*. 495/95 Express Lanes, On the Road Now. https://www.expresslanes.com/on-the-road. Accessed July 1, 2017.
23. *GitHub, Inc*. Managed and Tolled Lanes Feed Specification. Santa Clara Valley Transportation Authority. https://github.com/vta/Managed-and-Tolled-Lanes-Feed-Specification. Accessed July 1, 2017.
24. *Washington State Department of Transportation*. Traveler Information API. http://wsdot.wa.gov/traffic/api/. Accessed July 1, 2017.
25Ettelman, B., C. Simek, and T. Geiselbrecht. *Traveler Information Survey*. Texas Department of Transportation Technical Memorandum, 2016.
26. Smith, A. *Record Shares of Americans Now Own Smartphones, Have Home Broadband*. Pew Research Center. http://www.pewresearch.org/fact-tank/2017/01/12/evolution-of-technology/. Accessed July 13, 2017.
27. Google. *GTFS Realtime Overview – Vehicle Position*. https://developers.google.com/transit/gtfs-realtime/guides/vehicle-positions . Accessed July 30, 2017.
28. Google. *Choose a Data Submission Method*. https://support.google.com/transitpartners/answer/2529132. Accessed July 30, 2017.

The Standing Committee on Managed Lanes (AHB35) peer-reviewed this paper (18-05200).

Article

TRR
JOURNAL OF THE TRANSPORTATION RESEARCH BOARD

Safety Evaluation of Cross-Sectional Elements of Freeway Managed Lanes

Transportation Research Record
2018, Vol. 2672(14) 20–28

DOI: 10.1177/0361198118755703
journals.sagepub.com/home/trr
SAGE

Raul E. Avelar[1] and Kay Fitzpatrick[1]

Abstract

This paper investigated the safety implications of managed-lane cross-sectional elements. The objective of the Federal Highway Administration project was to investigate the relationship between crashes and buffer-separated managed lane dimensions. The results from several previous research studies have demonstrated that reductions in freeway lane width or shoulder width are associated with more crashes. This research found that a narrower managed-lane envelope width (i.e., left shoulder, managed-lane, and buffer width combined) is also associated with more freeway crashes for crashes of all severity levels including fatal and injury. Wider envelopes are associated a reduction in total freeway crashes, of all severities, of 2.8% in Texas, and 2.0% in California, for each additional foot of envelope width. In California, wider envelopes are associated with a reduction of 4.4% in managed-lane-related crashes resulting in fatalities and injuries at all severity levels, for each additional foot of envelope width. The analysis was conducted on nonweaving managed-lane segments that included a single managed-lane separated from the general-purpose lanes with a flush buffer area. The dataset included crashes on 128.0 miles of freeway in California (all 128.0 miles with flush buffers) and 60.4 miles of freeway in Texas (41.7 miles with pylon buffers and 18.7 miles with flush buffers). The California sites included freeways with three or four general-purpose lanes, and the Texas freeways had three to five general-purpose lanes.

Managed lanes (MLs) are "designated (also defined as preferential) lanes and roadway facilities located on or adjacent to controlled-access urban/suburban metropolitan highways that are actively operated and managed to preserve operational performance over comparable general traffic lanes" (*1*). Such preferential service often implies faster travel speeds and better reliability than on adjacent general-purpose (GP) lanes. Buffer separation and other geometric strategies can help to preserve these benefits. Recent research has found that higher speeds are associated with wider buffers (*2*). The wider ML buffer separation, travel lane, or shoulder width may also positively influence safety.

Research Objectives

To investigate the potential safety benefits, the Federal Highway Administration (FHWA) sponsored research (*3*) to determine the relationship between ML cross-section elements and crashes to evaluate the safety implications. This research examines different widths of the three key ML geometric elements: lane, buffer, and inner and outer shoulder. Access to and from a ML is assumed to affect safety so this research focused on the effects of the cross-section elements within a segment. Future research should attempt to quantify the relationship between crashes and ML access characteristics.

Literature Review

This section presents a summary of recent research for safety on freeways and on freeways with MLs.

Safety on Freeways

The *Highway Safety Manual* (HSM) includes crash prediction methods for freeways (*4*), and the research conducted as part of National Cooperative Highway Research Program (NCHRP) 17-45 formed the basis of the material in those chapters (*5*). The researchers found that reductions in lane widths and inside (left) shoulder widths are associated with increased crashes. The range of shoulder widths included in the NCHRP 17-45 study was 2–12 ft. An inside shoulder width of 6 ft was assumed as the base condition (*6*).

A recent Texas Department of Transportation (TxDOT) project examined the trade-offs of reducing lane and shoulder widths to permit an additional freeway lane and identified increased crashes when the widths of lanes or shoulders

[1]Texas A&M Transportation Institute, College Station, TX

Corresponding Author:
Address correspondence to Raul E. Avelar: r-avelar@tamu.edu

are reduced (*7*). Some identified safety associations included the following:

- Freeway 12-ft lanes tend to experience fewer fatal and serious injury crashes than 11-ft lanes, depending on the number of lanes in the cross-section: from 5% fewer fatal and serious injury crashes on four-lane freeways and up to 12% fewer on 10-lane freeways.
- Freeways with increased left shoulder width tend to experience 5% fewer crashes per additional foot of left shoulder, when there are no changes to the other model variables.
- Freeways with wider right shoulders tend to experience 9% fewer serious and fatal crashes per additional foot of right shoulder, when there are no changes to the other model variables.
- It was estimated that between 24% and 26% fewer fatal and serious injury crashes are expected for each additional lane, depending on the average lane width of the cross-section (within the lane width range of 11–12 ft) other model variables equal.

Whereas the research identified that an additional lane can result in reductions in crashes, whether the benefits of the additional lane completely offset the consequences of the reduced lane and shoulder widths would depend upon the conditions present at the site (*7*). The TxDOT study focused on freeways with GP lanes rather than freeways that include a ML.

ML or High-Occupancy Vehicle Lanes

Using crash data from years 2005 to 2007 at 13 southern California segments, totaling 153 miles, a 2013 paper reported on the relationship between cross-section design (i.e., lane width, shoulder width, and buffer width) and safety performance for high-occupancy vehicle (HOV) lanes (*8*). The segments had HOV lanes buffer-separated from the GP lanes. The analysis included crashes that occurred on the median shoulder, in the HOV lane, or in the adjacent GP left lane. Independent variables included geometric attributes and annual average daily traffic (AADT). The authors found that wider HOV lanes tended to be associated with lower crash frequencies, except for the 13-ft lanes. The research concluded that higher AADTs in HOV and left lanes tend to have higher crash frequencies, for all studied crash types, except injury crashes in the left lane. It was also found that a wider shoulder was associated with fewer HOV lane crashes. Statistical evidence was weak regarding buffer and left lane widths, so the research could not reject the null hypothesis of no safety effect for these geometric features.

Cothron et al. and Cooner and Ranft reported on the safety issues and impacts associated with buffer-separated concurrent-flow HOV lanes in two urban freeways in Dallas, Texas (*9,10*). They reviewed hard-copy crash reports for multiple years. The researchers identified the following factors associating with increased injury crash rates:

- High daily traffic volumes and extensive congestion in the GP lanes.
- Ramp–pair combinations at or near the minimum ramp terminal spacing as recommended by the American Association of State Highway and Transportation Officials (AASHTO) in *A Policy on Geometric Design of Highways and Streets* (commonly known as the *Green Book*) (*11*).
- Reduced HOV cross-section.
- Speed differential between the HOV and adjacent GP lane traffic.

An empirical-Bayes before–after study was conducted in Minnesota by Cao et al. to evaluate the effects of tolling on the I-394 MnPass Lanes, which opened for operation in 2005. AADT data from 1998 to 2008 were used (*12*). Crash data were obtained from the Minnesota Department of Transportation and other Twin-Cities metropolitan agencies. The study found a reduction of 5.3% in the overall number of crashes, with an economic benefit of $5 million from 2006 to 2008. However, the authors showed concern about transferability of the results to other HOT lane projects.

A 2012 paper presented a safety analysis evaluating time-of-day strategy for ML that concurrently allows use of the inner left lanes by HOVs and use of right shoulders as GP lanes during peak hours over a 3-year period, identifying lane specific AADT volumes, including the inner left lanes for HOV-only use during peak hours, GP lanes, right shoulder lanes, and all lanes as a whole (*13*). Negative binomial regression models were fitted to the data, including the AADT across all lanes, and no evidence found that the interest factors, including the ML strategy during peak hours, were associated with changes in crash frequency. The variable AADT volumes in the specific analysis of GP lanes showed about a 2% increase in weekday crashes for each increase of 1000 vehicles per day, in the AADT range of 50,000 to 83,000 vehicles per day. Right shoulder-specific analysis showed an increase of about 38% in crashes at the merge and diverge areas during adverse light conditions. This study concluded that the ML strategy does not appear to relate to changes in crash frequency significantly at the inner left lanes for HOV, GP lanes, or right shoulders.

Srinivasan et al. developed crash prediction models for freeway facilities with HOV and high-occupancy toll (HOT) lanes. Models were developed for 6-, 8-, 10-, and 12-lane freeways, reflecting both directions and including MLs (*14*). For most of the models, the left shoulder width was the most significant variable, other than segment length and AADT. An increase in left shoulder width was associated with a decrease in crashes. The effect of buffer type on crashes was statistically significant only in the model for 10-lane

Table 1. Description of Candidate Variables

Variable name	Description
AADT	Annual average daily traffic for the freeway (vehicles/day)
AADTHV	Annual average daily traffic for the ML (vehicles/day)
AADTMainL	Annual average daily traffic for the GP lanes (vehicles/day)
Buf_Type	Buffer type between ML and GP freeway lanes, either pylons or flush
Buf_W	Buffer width (ft)
GP_Trvl_W	GP lanes, travel width for GP lanes, determined as number of lanes multiplied by average lane width (ft)
GP_Adj_W	GP lanes, width of lane adjacent to the ML (ft)
GP_All_Ln_W	GP lanes, width of all GP lanes (ft)
GP_Avg_Ln_W	GP lanes, average lane width (ft)
GP_Ent	GP lanes, number of entrance ramps within the segment
GP_Exit	GP lanes, number of exit ramps within the segment
GP_NumLn	GP lanes, number of GP lanes that are not barrier separated and are moving in same direction
GP_R_Shld_W	GP lanes, right shoulder width (ft)
GP_Weave	GP lanes, number of weaving areas within the segment
ML_Env	ML envelope, sum of left shoulder width, lane width, and buffer width (ft)
ML_L_Shld_W	ML, left shoulder width (ft)
ML_Ln_W	ML, lane width (ft)
PSL	Posted speed limit (miles per hour)
T_Trvl_W	Total travel width (ft)

freeways, with an inclusion of a 2-ft to 3-ft buffer, with fewer fatal and injury crashes.

Data Collection and Preparation

This section describes the selection of sites, data elements to be collected, and data collection activities for this research.

Potential Variables for Study

The research team considered the findings from the literature and developed a list of potential variables that could be potentially influential on safety for MLs within a freeway corridor. Table 1 provides descriptions of the geometric variables selected for this study.

Site Selection

The first step in site selection was to identify the cross-section elements for study. The lane and shoulder width were obvious cross-sectional elements. Buffer separation rather than barrier separation was the focus for this research, considering (*a*) areas with pavement markings only, called flush buffers or (*b*) areas with pavement markings and supplemental pylons, called pylon buffers. The following were additional decisions made during this stage of the study:

- Focus on sites with one (rather than two) ML(s) per direction;
- Eliminate sites that have reversible operations;
- Select sites where the ML is operational 24 hours per day, 7 days per week;
- Seek sites that represent a range of buffer widths.

For this evaluation, efforts were focused on identifying potential sites in the three highway safety information system states with MLs (California, Minnesota, and Washington), and in Texas. Based upon the initial review of variables available within these crash databases, the states of California and Texas were selected for the study.

Highway Safety Information System Crash Data for California

At the time of data collection, the most recent 5-year data available for California was 2007 to 2011. Data were requested from selected counties in the Los Angeles area. It was initially determined that there were 15,257 crashes with the HOV code and 437 crashes with the HOV buffer code. The California study locations were on I-105, SR 134, I-210, and I-405.

Texas Data

The Texas Crash Records Information System was utilized for variables that could be used to identify HOV-specific crashes. Texas data offered additional details in extended fields (e.g., longitude, latitude). The latitude and longitude information allowed the research team to quickly merge the crash and additional geometric data collected for specific sites. Owing to the limited number of HOV-related crashes, the evaluation of the Texas data was limited to using all freeway crashes.

Five Texas freeways were selected: I-635 and US 75 in Dallas; and US 290, I-10, and US 59S in Houston. The Texas locations reflect both pylons and flush buffer segments.

Both California and Texas Data

Both Texas and California provided a mix of buffer widths and lane widths. The project requirements were to include a minimum of 12 sites. The research team identified 18 corridors and nine freeways with each direction uniquely considered, in case some corridors had to be eliminated due to unexpected challenges with the crash data.

Defining Segments

Within the identified corridors, segments were defined by the location of ML access control change. A new segment would start when access was (or was not) permitted. Each segment was then defined as being ML weaving, ramp, or nonweaving segment. For this study, only nonweaving segments were included in the analysis. Sites with three and four GP lane freeways in California and with three to five GP lanes in Texas were included. The minimum length of a segment was defined as 0.11 miles. Most California sites were between 0.8 and 1.7 miles and between 0.9 and 1.4 miles for Texas. The assembled roadway database consisted of 128.0 miles in California (all 128.0 miles with flush buffers) and 60.4 miles in Texas (41.7 miles with pylon buffers and 18.7 miles with flush buffers), after removing segments where weaving was expected, locations undergoing construction, and locations with no AADT data available.

Variables

For each segment identified, the research team collected geometric characteristics using Google Earth. As this package allows the user to compare satellite images taken at different points in time, it was possible to annotate the date of the earliest satellite image containing a matching set of ML characteristics to the most current image. In almost all cases, the change reflected when the ML was added to the freeway. This was used with the purpose of excluding any periods when the ML geometric characteristics were not consistent.

Table 1 provides descriptions of the geometric variables for the analyses. The research team gathered the information for the geometric variables by using the measurement tool available in Google Earth including the posted speed limit information by using the StreetView feature available in the Google Earth suite of tools.

Dataset Characteristics: Geometrics

Table 2 provides summary statistics of the key elements considered in this research. All the buffers for the California segments were flush, that is, no pylons with widths that varied between 1 and 12 ft. For this study, the California data reflected 2007–2011 and the buffers were separated by white and yellow lane line markings. Note that, since 2012, California began to use all white lane buffer markings when they refresh their markings.

The larger buffer widths (9 or 12 ft) included in the study were associated with preserving space for a downstream ML ramp on I-405, a route where buffer widths were otherwise narrow. Several freeways have buffers with a 4–5 ft width as illustrated in Figure 1.

The buffers in Texas include flush buffers, ranging from 1.5 to 5.0 ft, and pylon buffers between 4.0 and 6.0 ft. The buffer pavement markings in Texas use white lines. Figure 2 shows an example of pylon buffer separation on Texas freeways.

Analysis and Results

Research Methods

Because more than one and unequal time periods from each site were used as the unit of analysis, an appropriate statistical methodology had to consider the grouping structure in the data explicitly. Generalized linear mixed models can account for such a structure, while performing the type of parametric estimation that would allow for answering the research questions. These models are constructed under the assumption that crashes at study sites for a period follow a given probability distribution. Addition details are available in the research report by Stanitsas et al. (*15*). Although most periods were 1 year long, the methodology allows the incorporation of partial years into the analysis, as this variable is handled as exposure in a way similar to segment length.

The exposure variables in the model is the product of the length of time (expressed in years) and the segment length (expressed in miles) for each record in the database. This quantity has units of mile-years. The interpretation of the result will be in terms of the change in expected yearly crashes per mile corresponding to changes in the critical observed variables. The next section review results from estimating the model from the methodology just described. All analyses were performed using open-source statistical packages and libraries (*16, 17*).

Modeling and Results

The pool of variables to develop models was selected based on the findings of previous studies, along with the variables that define exposure. For the GP lanes, the number of lanes, lane width (average and per lane values), right shoulder width, number of entrance ramps, and number of exit ramps

Table 2. Range of ML Envelope Geometric Data by Corridor

Hwy	Dir	F or P	NW Length (mi)	SW-Min	SW-Avg	SW-Max	LW-Min	LW-Avg	LW-Max	BW-Min	BW-Avg	BW-Max
CA 105	EB	F	9.44	8.5	10.8	13.0	10.5	10.9	11.5	4.5	4.8	5.0
CA 105	WB	F	13.39	8.0	10.7	20.0	11.0	11.6	12.0	5.0	5.0	5.0
CA 134	EB	F	8.07	1.0	3.5	15.0	10.8	11.2	12.0	1.5	1.5	1.5
CA 134	WB	F	7.55	1.0	1.3	2.0	11.0	11.2	11.5	1.0	1.6	2.0
CA 210	EB	F	19.13	1.0	7.0	20.0	11.0	11.3	12.0	2.5	3.2	5.0
CA 210	WB	F	14.16	1.0	7.9	20.0	11.0	11.4	12.0	2.5	3.4	5.0
CA 405	NB	F	29.7	1.0	4.0	33.0	10.0	10.7	11.5	1.0	2.6	12.0
CA 405	SB	F	26.56	1.0	4.1	21.0	10.0	11.1	12.0	1.0	3.5	12.0
TX 75	NB	P	11.0	3.0	3.2	3.5	11.0	11.2	11.5	4.0	4.0	4.0
TX 75	SB	P	11.0	2.0	2.0	2.0	11.5	11.5	11.5	4.0	4.0	4.0
TX 635	EB	P	8.1	2.0	2.6	3.5	10.0	10.1	10.5	4.0	5.4	6.0
TX 635	WB	P	7.4	1.0	1.5	2.5	10.0	10.4	10.5	5.5	5.5	5.5
TX 10	EB	P	2.3	17.5	17.8	18.0	13.0	13.3	13.5	5.0	5.3	5.5
TX 10	WB	P	1.9	17.5	17.5	17.5	12.5	12.5	12.5	5.5	5.5	5.5
TX 59	NB	F	7.3	10.0	11.8	13.0	11.0	11.4	12.0	1.5	3.7	5.0
TX 59	SB	F	6.0	9.0	10.6	12.0	11.0	11.8	12.0	2.0	3.2	5.0
TX 290	NB	F	2.2	1.5	2.3	4.0	10.5	10.8	11.5	1.5	1.5	1.5
TX 290	SB	F	3.2	1.5	1.5	1.5	11.0	11.0	11.0	2.0	2.0	2.0

Note: Hwy = State and highway number; Dir = direction, where NB = northbound, SB = southbound, EB = eastbound, and WB = westbound; F or P: F = flush buffer and P = pylons present within buffer area; NW Length = sum of the lengths for nonweaving segments within the corridor; SW = shoulder width (ft); LW = lane width (ft); and BW = buffer width (ft); Min = minimum width for the highway direction; Avg = average width for the highway direction; Max = maximum width for the highway direction.

Figure 1. Examples of wide buffers on California I-405.

within the segment were investigated. For ML components, models were considered that treated these elements as unique variables (e.g., buffer width, ML width, left shoulder width) or as a combined unit (e.g., shoulder, lane, and buffer widths added into a ML envelope width variable). This combination of variables intended to mitigate the inherent correlation between the variables. In the dataset for this research, the widths of left shoulder, ML, and buffer all tended to vary concurrently: at sites where one of these variables is wide, the other two variables tended to be wide as well.

All Freeway Crash Frequency: California

For California, the modeling process began with a full model including all the key cross-section variables. Only three coefficients showed statistical significance at this stage: the

Figure 2. Example of pylons in buffer on Texas freeways.

Table 3. Total Cross-Section Crash Frequency Model (All Severities) in California

Variable	Estimate	Standard error	z value	Pr(>|z|)[a]	Significance[b]
(Intercept)	−12.442053	2.666415	−4.666000	0.000003	***
log(AADT/2)	1.397905	0.225244	6.206000	<0.000001	***
ML_Envelope	−0.020306	0.005014	−4.050000	0.000051	***

[a]*p*-values reported.
[b]Significance values are as follows: *** $p < 0.001$.

intercept, the exponent for AADT, and the total ML envelope width. However, this model is a reference point to observe the trends of each variable, after accounting for the effects of other cross-sectional variables. A similar trend to that of the ML envelope was observed for the width of the rest of freeway cross-section, but not found to be statistically significant, along with the coefficients for other variables of interest. When the ML envelope components were considered separately, ML width became the only significant variable in addition to AADT. The correlation between buffer and left shoulder width, as typically wider buffers found where wider left shoulders are present, it was believed to be affecting the significance of those estimates, and potentially biasing others in the model.

A reduced model derived from the full model initially used was fitted to exclude variables that are not significantly related to total crashes. Table 3 shows the results and the results of interest are practically unchanged: a reduction of 2.0% in crashes per additional foot of envelope width (1 − exp(−0.020306)).

Proportion of ML Crashes to All Crashes: California

To continue to explore the potential influence of buffer width on total crashes, a model was evaluated where the proportion of ML crashes to total crashes was the response variable. It is expected that this proportion, modeled as a binomial variable, may capture safety differences associated with differences in the freeway cross-section elements. The residuals in the initial results showed moderate over-dispersion. This condition affects the standard error of the regression estimates and must be considered. The log-likelihood of the model was adjusted to represent a quasi-binomial distribution, that is, explicitly correcting for extra-binomial dispersion.

The coefficients in the full model indicated that both the width of the ML and the width of the buffer are significantly related to the proportion of ML crashes, with the lane width being the more influential factor. Other factors with statistically significant associations include AADTMainL and AADTHV, but several GP related variables were not found to be statistically significant. A model reduction effort was carried out, arriving at the model in the middle column of Table 4, which was most parsimonious for crashes of all severities. The magnitudes and directions of the results were virtually unchanged compared to the full model.

The interpretation of these results for crashes of all severities is as follows: given that a crash has occurred on a freeway containing a single ML and a flush buffer, the odds of that crash being related to the ML or the buffer decrease of 23% with each additional foot of width in the ML, or equivalently

Table 4. California Model for Proportion of ML to Freeway Crashes (Models by Severity Levels)

Severity of crashes	Crashes of all severities	Fatal and injury crashes
Variable	Estimate (SE) significance[a]	Estimate (SE) significance[a]
(Intercept)	0.3272 (0.8673)	−0.9624 (0.249)***
AADTMainL	-1.00×10^{-5} (9.00×10^{-6})**	na
AADTHV	4.00×10^{-5} (1.00×10^{-5})***	na
ML_Ln_W	−0.2058 (0.0724)**	na
Buf_W	−0.1166 (0.024)***	na
GP_Ent	0.0809 (0.0367)*	na
ML_Envelope	na	−0.0331 (0.0129)*

Note: na = not applicable; SE = standard error.
[a]Significance values are as follows: ***$p < 0.001$; **$p < 0.01$; *$p < 0.05$; and no symbol = not significant.

Table 5. California, Reduced Model for ML-Related Crashes (Models by Severity).

Severity of crashes	Crashes of all severities	Fatal and injury crashes
Variable	Estimate (SE) significance[a]	Estimate (SE) significance[a]
(Intercept)	1.1378 (1.8911)	−3.1824 (2.2628)
log(AADTHV)	0.5013 (0.1465)***	0.3549 (0.241)
ML_L_Shld_W	−0.0372 (0.0146)*	na
ML_Ln_W	−0.3915 (0.1063)***	na
Buf_W	−0.0772 (0.0456)#	na
ML_Env	na	−0.0447 (0.0127)***

Note: na = not applicable; SE = standard error.
[a]Significance values are as follows: ***$p < 0.001$; *$p < 0.05$; #$p < 0.10$; and no symbol = not significant.

by a factor of 0.77 (exp(−0.2058)). Similarly, the odds of a crash being related to the ML or buffer decrease by 11% with each additional foot of width in the flush buffer, or equivalently by a factor of 0.89 (exp(−0.1166)).

Fatal and Injury Proportion of ML Crashes to All Freeway Crashes: California

Similar to all severity levels, the relationship of the ML envelope with fatal and injury crashes was examined and found the association to be negative. After undergoing model selection from a full model, a reduced model was applied with result shown in the last column of Table 4. ML envelope proved the only statistically significant term in this model. When the envelope is separated into left shoulder width, lane width, and buffer width, none of the widths are significant. The research team believes that this is due to the significantly smaller number of crashes involved in the model, compared with the model in the middle column of Table 4 that represents crashes of all severities.

ML-Related Crash Frequency: California

All Severity Levels Crashes. An evaluation was also performed using only crashes that were coded as being on the ML or on the buffer. The final reduced model for all severity levels, only including the significant variables, is shown in the middle column of Table 5, after model selection. All components of the ML envelope, left shoulder width, lane width, and buffer width, are significant, along with the volume in the ML (AADTHV). The coefficients indicate that the lane width is the most influential variable followed by the buffer width. Per this model, changes in left shoulder width are not as influential in the number of ML-related crashes as changes in the buffer or lane width.

Fatal and Injury Crashes. When considering only fatal and injury severity level ML-related crashes, a full model showed roughly the same trends as in the middle column of Table 5 but, given the reduced number of crashes, the statistical significance of those estimates was reduced in comparison. The modeling effort did not arrive to a meaningful reduced model when considering the ML elements separately. Model selection was utilized considering the ML elements as an envelope, similar to Tables 3 and 4 and shown in last column of Table 5. Results indicate that each additional foot of envelope is associated with a 4.4% reduction in ML and buffer-related crashes (1 − exp(−0.04471)).

An interesting observation from a comparison between the two models in Table 5 is the different rate of increase of ML crashes by traffic volume. For example, a 20% increase in AADTHV corresponds to an increase of 9.6% in crashes of all severities (i.e., $1.20^{0.5013} = 1.096$). The same increase of

Table 6. Reduced Model for Total Freeway Crash Frequency (All Severity Levels) in Texas

Variable	Estimate	Standard error	z value	Pr(>|z|)[a]	Significance[b]
(Intercept)	0.42185	1.45744	0.289	0.77224	
log(AADT/2)	0.23482	0.12755	1.841	0.06563	#
ML_Env	−0.02808	0.01603	−1.752	0.07979	#
Buf_Type=Pylons	0.66049	0.22595	2.923	0.00346	**

[a]*p*-values reported.
[b]Significance values are as follows: blank cell = not significant; $^{\#}p < 0.10$; $^{**}p < 0.01$.

20% in AADTHV is associated with only an increase of 6.7% in fatal and injury crashes (i.e., $1.20^{0.3549}$=1.067). Therefore, the analysis indicates that a higher proportion of fatal and injury crashes tend to occur at lower ML volumes compared to higher ML volumes.

All Freeway Crash Frequency: Texas

All Severity Levels Crashes. Several variables were utilized for the initial model of total Texas freeway crashes, at all severity levels. Some variables, such as number of GP lanes, were expected to be significant in this analysis but were not, or the results were counterintuitive. It was expected that the number of crashes would increase rather than decrease as the number of GP entrance ramps increased however, this relationship was not significant. It is suspected that this was because the segments were created to focus on ML segments with no access openings, so the relationship of crashes with the characteristics present on the neighboring GP lanes may not represent enough variation to reliably estimate the coefficients. Additional exploration would be needed to explain the potential relationships between crashes and those roadway geometric variables, which was beyond the scope of this project.

After model selection, most nonsignificant variables drop from the model, arriving at the results shown in Table 6. Fewer crashes are expected for wider ML envelopes. A reduction of 2.8% in crashes is associated with each additional foot of envelope width (1 − exp(−0.02808)).

A notable finding in this table is the magnitude and sign of the coefficient for buffers including pylons. According to this model, sites with pylon buffer separation tend to have significantly more crashes than sites with flush buffers. Perhaps this is because the pylons represent an obstacle that increases the opportunities for crashes to occur. If that were so, one may expect the increase in crashes to exclude higher-severity outcomes. In any case, more research is needed to better understand the implications of using pylons to separate the ML from the GP lanes.

Conclusions

This study has investigated the relationship between crashes and buffer-separated ML dimensions. Facilities in this study include MLs on freeways separated by a buffer and located to the left of the GP lanes.

The crash frequency analyses of the Texas and California data showed that wider ML envelope (i.e., left shoulder, ML, and buffer) widths are associated with fewer freeway crashes when considering all severity levels as well as when considering only fatal and injury severity levels. Specific findings included the following:

- In Texas, wider envelopes are associated with a reduction of 2.8% in total freeway crashes (all severities) for each additional foot of envelope width.
- In California, wider envelopes are associated with a reduction of 2.0% in total freeway crashes (all severities) for each additional foot of envelope width.
- In California, wider envelopes are associated with a reduction of 4.4% in ML-related crashes (fatal and injury severity levels) for each additional foot of envelope width.

When focusing on the envelope components, that is, shoulder, lane, or buffer, trends in the data clearly suggest that fewer crashes are associated with wider buffer widths. However, an attempt to quantify this trend resulted in strongly correlated estimates.

When exploring whether a particular component of the ML envelope is more influential than another, the simultaneous evaluation on the three envelope components using California freeway crashes, at all severity levels, identified the left shoulder width as the only statistically significant association with a reduced number of crashes.

When modeling variations in the proportion of California ML-related crashes to all crashes including ML and GP crashes, this research found associations with some key variables. For example, the probability of a crash being a ML-related crash:

- Decreases with increasing volume in the GP lanes (p = 0.0034 for a 0.99 odds ratio estimated per each additional 1000 vehicles per day; 0.99 = exp[−0.00001 × 1000]).
- Increases with increasing volume in the ML ($p < 1 \times 10^{-5}$ for a 1.04 odds ratio estimated per each additional 1000 vehicles per day; 1.04 = exp[−0.00004 × 1000]).

- Decreases with increasing ML width ($p = 5.3 \times 10^{-3}$ for a 0.814 odds ratio estimated per each additional foot of ML width; $0.814 = \exp[-0.20576]$).
- Decreases with increasing buffer width ($p < 1 \times 10^{-5}$ for a 0.89 odds ratio estimated per each additional foot of buffer width; $0.89 = \exp[-0.11657]$).
- Increases with increasing number of entrance ramps ($p = 0.0294$ for a 1.08 odds ratio estimated per each additional GP entrance ramp within the segment; $1.08 = \exp[0.08092]$).

An evaluation was also performed on the crashes that were coded as being on the ML or in the buffer. The refined model on ML and buffer-related crashes in California that only includes the significant variables found that all three components of the ML envelope (left shoulder width, lane width, and buffer width) are significant along with the volume in the ML. The results indicate that the lane width is the most influential element, followed by the buffer width. Changes in left shoulder width were not found to be as influential in the number of ML or buffer-related crashes as the buffer or lane width changes.

The findings from this research are that wider ML envelopes are associated with fewer crashes and pylons appear to be associated with more crashes. The dataset used to identify the pylon crash relationship presented two important challenges: dataset size and use of freeway, rather than ML-related, crashes. If only those crashes near the pylons are considered, say for the lanes on either side of the pylons, results may be different. The available data for this study were limited in total number of miles and only one state had data with pylons. The benefits or disadvantages of using pylons, and the available buffer width present between the travel lane and the pylon need additional research.

Acknowledgments

This paper is based on research sponsored by FHWA under the supervision of Greg Jones. The research was performed at the Texas A&M Transportation Institute. The authors thank the large number of undergraduate and graduate student workers who helped with data reduction. We also thank the project panel and reviewers of earlier versions of this work for their feedback during the course of this study.

References

1. Fitzpatrick, K., M. Brewer, S. Chrysler, N. Wood, B. Kuhn, G. Goodin, D. Ungemah, V. Dewey, B. Perez, N. Thompson, C. Swenson, and D. Henderson. *Guidelines for Implementing Managed Lanes*. Transportation Research Board, 2016.
2. Fitzpatrick, K., R. Avelar, and T. Lindheimer. Operating Speed on a Buffer-Separated Managed Lane. *Transportation Research Record: Journal of the Transportation Research Board*, 2016. 2616: 1–9.
3. Fitzpatrick, K., and R. Avelar. *Safety Implications of Managed Lane Cross Sectional Elements*. FHWA, Washington, D.C., 2016.
4. AASHTO. *Highway Safety Manual Supplement*. American Association of State Highway and Transportation Officials, Washington, D.C., 2014.
5. Bonneson, J. .A, S. Geedipally, M. Pratt, and D. Lord. *Safety Prediction Methodology and Analysis Tool for Freeways and Interchanges*. Transportation Research Board, Washington, D.C., 2012.
6. Miller, C., R. Deuser, J. Wattleworth, and C. Wallace. *Safety Evaluation of Priority Techniques for High-Occupancy Vehicles*, Federal Highway Administration, Washington, D.C., 1979.
7. Dixon, K., K. Fitzpatrick, R. Avelar, M. Perez, S.E. Ranft, R. Stevens, S. P. Venglar, and A. P. Voigt. *Reducing Lane and Shoulder Width to Permit an Additional Lane on a Freeway*, Texas Department of Transportation, Austin, TX, 2015.
8. Jang, K., S. Kang, J. Seo, and C. Y. Chan. Cross Section Design for the Safety Performance of Buffer-Separated-High-Occupancy Vehicle Lanes. *ASCE Journal of Transportation Engineering*, No. 139, 2013, pp. 247-254.
9. Cothron, S.A., S. E., Walters, C. H. Ranft, D. W. Fenno, and D. Lord. *Crash Analysis of Selected High-Occupancy Vehicle Facilities in Texas: Methodology, Findings and Recommendations*. Texas Transportation Institute, Texas A&M University System, College Station, TX, 2005.
10. Cooner, S. A, and S. E. Ranft. Safety Evaluation of Buffer-Separated High-Occupancy Vehicle Lanes in Texas. *Transportation Research Record: Journal of the Transportation Research Board*, 2006. 1959: 168-177.
11. AASHTO. *A Policy on Geometric Design of Highways and Streets*. American Association of State Highway and Transportation Officials, Washington, D.C., 2001.
12. Cao, X., Z. Xu, and A. Huang. Safety Benefits of Converting HOV Lanes to HOT Lanes: Case Study of the I-394 MnPASS. *ITE Journal*, 2012, pp. 32-37.
13. Lee, J., R. Dittberner, and H. Sripathi. Safety Impacts of Freeway Managed-Lane Strategy: Inside Lane for High-Occupancy Vehicle Use and Right Shoulder Lane as Travel Lane During Peak Periods. *Transportation Research Record: Journal of the Transportation Research Board*, 2007. 2012: 113–120.
14. Srinivasan, S., P. Haas, P. Alluri, A. Gan, and J. Bonneson. *Crash Prediction Method for Freeway Facilities with High Occupancy Vehicle (HOV) and High Occupancy Toll (HOT) Lanes*. Florida Department of Transportation, 2015.
15. Stanitsas, P., J. Hourdos, and S. Zitzow. *Evaluation of the Effect of MnPASS Lane Design on Mobility and Safety*, Minnesota Department of Transportation, 2014.
16. The R Development Core Team. *R: A Language and Environment for Statistical Computing*. [Online] R Foundation for Statistical Computing, Vienna, Austria, 2013. Version 2.10.1 (2009-12-14). http://www.R-project.org. ISBN 3-900051-07-0.
17. Pinheiro, J. and D. Bates. *Mixed-Effects Models in S and S-PLUS*. New York: Springer, 2000. ISBN 0-387-98857-9.

The Standing Committee on Managed Lanes (AHB35) peer-reviewed this paper (18-00323) and selected it to receive its 2018 Best Paper Award.

Article

TRR
JOURNAL OF THE TRANSPORTATION RESEARCH BOARD

Transportation Research Record
2018, Vol. 2672(14) 29–38
© National Academy of Sciences:
Transportation Research Board 2018
Article reuse guidelines:
sagepub.com/journals-permissions
DOI: 10.1177/0361198118798296
journals.sagepub.com/home/trr
SAGE

Reliability Assessment Tool: Development and Prototype Testing

George F. List[1], Nagui Rouphail[1], Russell Smith[1], and Billy Williams[1]

Abstract

This paper presents a monitoring system that was developed to assess travel time reliability for observed operating conditions by utilizing traffic stream and non-transportation related data. A prototype was created for an interstate highway route in the Research Triangle region of North Carolina. It describes how the input datasets were obtained, the required data fusion procedures, how the data were analyzed to create the monitoring system outputs, and relevant insights obtained from the reliability monitoring system prototype.

Travel time reliability has become a major focus for the transportation sector as both researchers and practitioners endeavor to monitor and enhance the performance of transportation systems. Reliability has become a key performance measure for transportation agencies in the United States (*1*). The Strategic Highway Research Program (SHRP 2) deemed it one of the four factors that needed to be addressed in decision-making on capacity expansion (*2*). In addition, one of the recent surface transportation laws, the Moving Ahead for Progress in the 21st Century Act (MAP-21) requires states to report performance outcome achievements (*3*, *4*).

This paper describes the development of a prototype travel time reliability monitoring system focused on an interstate highway segment along I-40 in the Research Triangle region of North Carolina. It describes how the input datasets were obtained, the required data fusion procedures, how these data were analyzed to create the monitoring system outputs, and relevant insights obtained from the reliability monitoring system prototype.

Previous Research

The history of using probe data for travel time monitoring appears to have started with the pioneering ADVANCE project in Chicago (*5*). Vehicles equipped with dead reckoning sensors and time stamp recorders were used to synthesize travel times. The focus was on estimating average travel times, not their distributions.

Project L02, sponsored under the auspices of SHRP 2, developed a procedural framework for a travel time reliability monitoring system (TTRMS) that could utilize data from and integrate into existing traffic monitoring systems. See List et al. (*6*). A block diagram shows the structure suggested for a TTRMS (Figure 1). The modules are shown as boxes, while the inputs and outputs are shown as circles.

The four suggested modules were as follows. The data manager assembles incoming information from traffic sensors, weather data feeds, incident reporting systems, and other sources, and places them in a database. The "other systems" module acquires external data for the data manager. The computational engine prepares "pictures" of the system's reliability. The report generator invokes analyses and assembles information used to make decisions about capacity investments, operational changes, etc.

A general finding from L02 was that reliability can be described effectively by focusing on probability density functions (PDFs) and cumulative density functions (CDFs). Further, these distributions should be developed for both travel times and travel rates (travel rates normalize the travel times based on distance). The PDFs are helpful for identifying multi-modality or the existence of multiple operating conditions within the data being examined consistent with, for example, Barkley et al. (*7*), Guo et al. (*8*), and Fraley and Raftery (*9*). The CDFs are helpful for comparing operating conditions or systems and determining whether reliability was improving. Other popular reliability metrics, such as the buffer index

[1]North Carolina State University, Raleigh, NC

Corresponding Author:
Address correspondence to Russell Smith: rcsmit14@ncsu.edu

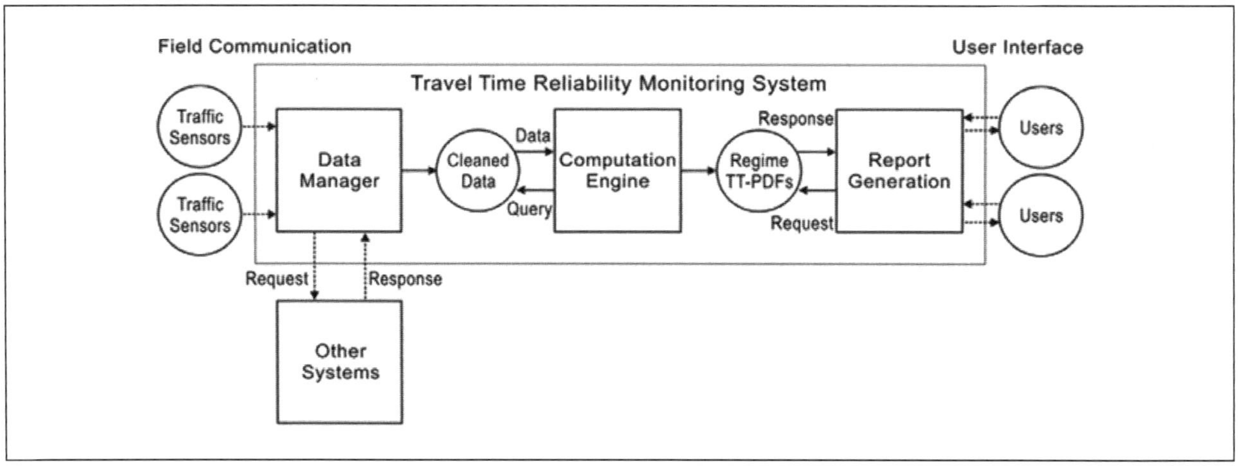

Figure 1. Travel time reliability monitoring system modules.

and travel time index, can be derived from these distributions. No aspect of reliability is overlooked.

Concepts presented in L02, as well as tools developed in companion SHRP 2 projects L07, L08, C11, and L05, were tested in pilot projects conducted in California, Minnesota, Florida, and Washington (*10*). The California site focused on two freeway facilities: I-210 in Los Angeles County and I-5 in Orange County. The Minnesota site focused on two freeways: I-94 from downtown Minneapolis to downtown St. Paul and I-35W from downtown Minneapolis south to Burnsville. The Florida site focused on the Dade County I-95 corridor general purpose lanes and express lanes in the northbound and southbound directions as well as the parallel SR7/US441 arterial. The Washington site focused on the I-5 and I-405 corridors from Lynnwood to Tukwila (approximately 30 miles long for each corridor running through the Puget Sound metropolitan region), and the SR-522 urban arterial near Seattle. All these pilots concluded that the SHRP 2 tools had the potential to monitor and model the reliability impacts and be incorporated into day-to-day processes, standard operating guidelines, and performance reporting systems.

Other research efforts have also focused on developing reliability monitoring systems. Among the original monitoring systems is the one developed to report congestion performance of urban freeway systems nationwide (*11*). More recently, monitoring systems have been developed for Pennsylvania Department of Transportation (PennDOT) (*12*) and Maryland DOT (*13*). Supporting analytical tools have been developed through the analysis of various data sources (*14–17*). Washington State DOT developed a web-based travel time estimator for public use based on the principle of travel time reliability (*18*).

Methodology

The methodology presented in the L02 report was followed to develop the prototype monitoring system. The year 2010 was selected as the study year due to the data available. The facilities examined were I-40, I-440, and I-540. The prototype uses a westbound I-40 route from US 1/US 64/I-440 to NC 147 (Durham Freeway) as a test facility.

Six interrelated datasets were developed: (i) travel rates, (ii) flow rates, (iii) incidents, (iv) maintenance activity, (v) weather, and (vi) planned events. Two primary keys were employed: time (which pertained to all six datasets; at a five-minute granularity) and Traffic Message Channel (TMC) segment (for the first four datasets).

For the travel rates, speeds by TMC segment were acquired from the VPP Suite (Vehicle Probe Project) Massive Data Downloader tool hosted by the CATT Lab's Regional Integrated Transportation Information System (RITIS) (*19*). The VPP Suite is a repository of traffic data for states participating in the I-95 Corridor Coalition Project. Speed observations were obtained for 203 TMC segments in total between the three freeways. The breakdown of TMC segments by facility was: 106 on I-40, 33 on I-540, and 64 on I-440. In the analysis year, 2010, Inrix was the data provider.

"Stitched" travel times for both TMC segments and routes were then developed from the raw speeds (*20*). These stitched values reflect average travel times that would have been experienced by vehicles that commenced their trips across the TMC segment or route within a five-minute interval. Travel times were developed for simulated vehicles entering the TMC or route every 30 seconds. The number of simulated trajectories used to calculate travel time can be adjusted as necessary. For the TMCs, the travel times were developed using

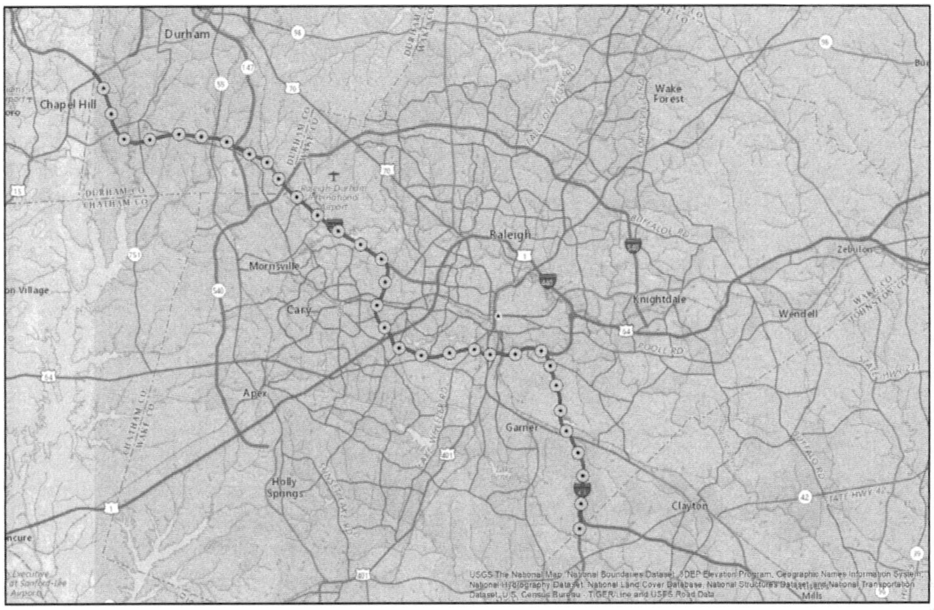

Figure 2. Study TMCs and traffic flow rate monitoring stations (*19, 21*).

travel rates from subsequent five-minute intervals if the temporal interval boundaries were crossed. For example, a simulated vehicle that began traversing the TMC at 4.5 minutes into the five-minute interval spent only a half minute traversing the TMC segment before the temporal interval boundary was reached. Assuming it was not yet at the end of the TMC segment (typically about a mile long), it would then begin using the travel rate for that same TMC from the next five-minute interval. This process would repeat until the end of the TMC segment was reached. In some instances, the subsequent travel rates were very like the previous ones, but in other situations they were quite different. Moreover, if the congestion level was high or there was an incident, the travel rates might be very slow, and values from more than one subsequent five-minute interval might have been used. The stitched travel times were then converted into stitched travel rates by dividing by the length of the TMC.

Similar ideas pertain to the stitched travel times for routes except that both spatial and temporal boundaries are crossed. For example, when the simulated vehicle reaches the downstream end of a TMC segment, it enters the next downstream segment and begins using the travel rate for that segment. Since the simulated vehicles each have a different starting time on the initial TMC segment in a five-minute time period, the points in time when they cross these TMC segment boundaries are all different. This means the stitched travel time for each simulated vehicle is uniquely different from the others.

Traffic flow data were also obtained from the RITIS Data Archive platform hosted by the CATT Lab (*21*) for the 120 HERE roadside side-fire radar units located in the Research Triangle region on I-40, I-440, and I-540 (Figure 2).Eight radar units were located along segments included in the prototype.

The flow rates were assigned to the TMCs as follows. For the TMC that corresponded to the sensor's location, the observed value was used directly. For TMCs that were upstream and downstream of that location but prior to the next interchange, that same flow rate was assigned. The flow rate was assumed to remain constant between interchanges. For the TMC segments at the interchanges, the average of the upstream and downstream observed values was used. To illustrate, if in a five-minute period the upstream sensor was reporting a flow rate of 5000 vph (across three lanes) and the downstream sensor was reporting 4000 vph, then the TMCs within the interchange would be assigned a value of 4500 vph.

The incident data was obtained from TIMS, the traffic incident management system used by North Carolina (NC) DOT. The primary purpose of TIMS is to provide information to the motoring public and includes unplanned incidents such as vehicle collisions and abnormally heavy congestion, and planned incidents such as a lane closures for road work. TIMS records indicate where the incident occurred (e.g., latitude/longitude and state mileposts), the nature of the incident, its duration, and in some cases, the number of lanes closed. Spot incidents such as collisions were geolocated by coordinates and segment incidents spanning several TMCs such as work zone lane closures were geolocated by state milepost. TIMS provides the statewide mile marker(s) for each incident, however, NCDOT roadway spatial data is referenced by county milepost. Therefore, the raw incident data were processed visually by plotting spot incidents and the location of state milepost signs as point features in GIS. Each incident was first assigned to a TMC(s) and five-minute interval(s) based on where the event occurred (the reported latitude/longitude) and the reported start and end time.

Speed data for upstream and downstream TMCs, during prior and subsequent five-minute time intervals were reviewed to assign the incident to time periods before and after the reported event and to TMCs upstream and downstream of the reported location if it appeared that the travel rates for those TMCs were adversely affected. Two baselines were used in these comparisons. One was the travel rate that should have pertained for the TMC if the operating conditions were "normal." The second was the contemporaneous travel rates on nearby TMCs before and after the five-minute time period being examined. This created an "impact area" for the incident in both space and time.

The weather data was obtained from a local weather monitoring station. The station chosen reported weather conditions every five minutes, which was very helpful. Each record reported the air temperature, the air temperature condition (e.g., freezing), and the precipitation condition (e.g., heavy precipitation). In Raleigh, where adverse conditions such as snow and heavy fog are rare, this level of specificity seemed sufficient. Significant adverse impacts from weather include severe thunderstorms and black ice.

Maintenance and construction information was obtained from TIMS. In some instances, the events pertain to only a few TMCs and are of short duration. In other cases, they cover many TMCs and are of long duration (e.g., the reconstruction of lanes along a significant portion of the freeway). The five-minute database was created manually. When completed, it showed maintenance activity for every TMC in each five-minute time interval.

The planned event database was developed by assembling information about sporting events, concerts, and other activities that took place during the analysis year. Affected five-minute time periods are associated with these events but to date, no specific TMCs have been assigned. The database management system has the potential to allow such assignments.

The analysis involved two steps. The first was to identify a spatial and temporal domain of interest. An example would be all the TMCs on a specific facility in a direction during the morning peak on weekdays. Another would be all the TMCs for the entire year. A third would be the TMCs along a route during abnormal conditions. The second step was to analyze the resulting pool of observations. This involved creating and analyzing histograms, cumulative histograms, PDFs, CDFs, composite PDFs and CDFs, and performance tables. Illustrations of each of these are provided in the next section.

Results

The analysis began with the selection of a spatial and temporal context. The spatial limits were a set of TMCs on I-40 westbound that are used heavily during the peak periods (Figure 2). The TMCs in sequence begin on the eastern end and progress to the west. The posted speed limit along the route is 65 mph. In addition, a single route was formed comprised of all the TMCs.

The first analysis studied the performance of the TMCs across the entire year. Weekdays were divided into a morning period from 6:00 to 10:00 a.m., a midday period from 10:00 a.m. to 3:00 p.m., and an evening period from 3:00 to 7:00 p.m. All other intervals were classified as "other." Observations on weekends (Saturday and Sunday) were assigned a separate category. Table 1 shows the distribution of travel rates (shown as speeds) that were observed for the TMCs during each time category.

Information such as that presented in Table 1 must be reported to FHWA in compliance with the MAP 21 monitoring requirements. Notice that on a 24-hour basis during weekdays, 90.9% of the time, the observed speeds are 60 mph or greater. Only 8.1% of the time were they at 60 mph or less. On the weekends, the performance is even better. However, during the morning peak, the speeds are lower; 81.3% of the time the speeds are 60 mph or greater, and 13% of the time (more than once every two weeks) they are 45 mph or less.

A deficiency of the assessment presented in Table 1 is that it does not examine travel times that would be experienced by vehicles that traverse the TMC segments. No route includes only a single TMC segment. In fact, a common route uses all twenty-three segments. It is instructive to examine the performance of this route (Table 1).

Notice that the performance of this route is significantly poorer than was the case for the TMC segments separately. Only 79.3% of the time is the speed of this route 60 mph or greater. 13.4% of the time it is 45 mph or less. Moreover, in the AM peak, only 51.1% of the time is it 60 mph or better. And 41.3% of the time it is 45 mph or less. Hence, if the policy performance of this

Table 1. Distribution of Speeds for the Examined TMCs and Route

Percentage distributions for the TMCs

	Weekdays					
Category	24-HR	AM	Midday	PM	Other	Sat/Sun
60 +	90.9	81.3	95.0	90.1	93.0	93.3
55–60	4.5	5.7	3.3	4.9	4.4	3.8
45–55	1.9	4.5	0.9	2.0	1.3	1.3
40–45	0.5	1.8	0.2	0.5	0.2	0.4
30–40	0.8	2.4	0.3	0.9	0.3	0.5
15–30	1.1	3.3	0.3	1.2	0.5	0.4
0–15	0.4	1.0	0.1	0.4	0.3	0.2

Percentage distributions for the route

	Weekdays					
Category	24-HR	AM	Midday	PM	Other	Sat/Sun
60 +	79.3	51.1	90.4	85.1	82.2	84.2
55–60	7.5	7.6	5.0	7.1	8.8	8.4
45–55	5.1	13.5	2.4	3.6	3.9	3.5
40–45	1.5	4.7	0.5	0.8	1.0	0.9
30–40	2.5	9.3	0.4	1.3	1.4	0.9
15–30	1.5	4.8	0.5	0.9	0.9	0.9
0–15	2.7	9.1	0.9	1.2	1.8	1.2

freeway is set to be 45 mph or better, then two out of every five weekdays that performance target would not be met. This route-based assessment is very important. It reveals problems that the TMC-based analysis misses.

Having identified the fact that the performance in the AM peak is the worst, the next question is why. Which are the problematic TMCs and what are the causal factors? Figure 3 shows the percentage of time that the TMCs are at 45 mph or less. Segments are numbered from east to west.

Clearly, the first 7 to 10 TMCs are the source of the trouble, particularly the first five. For example, the

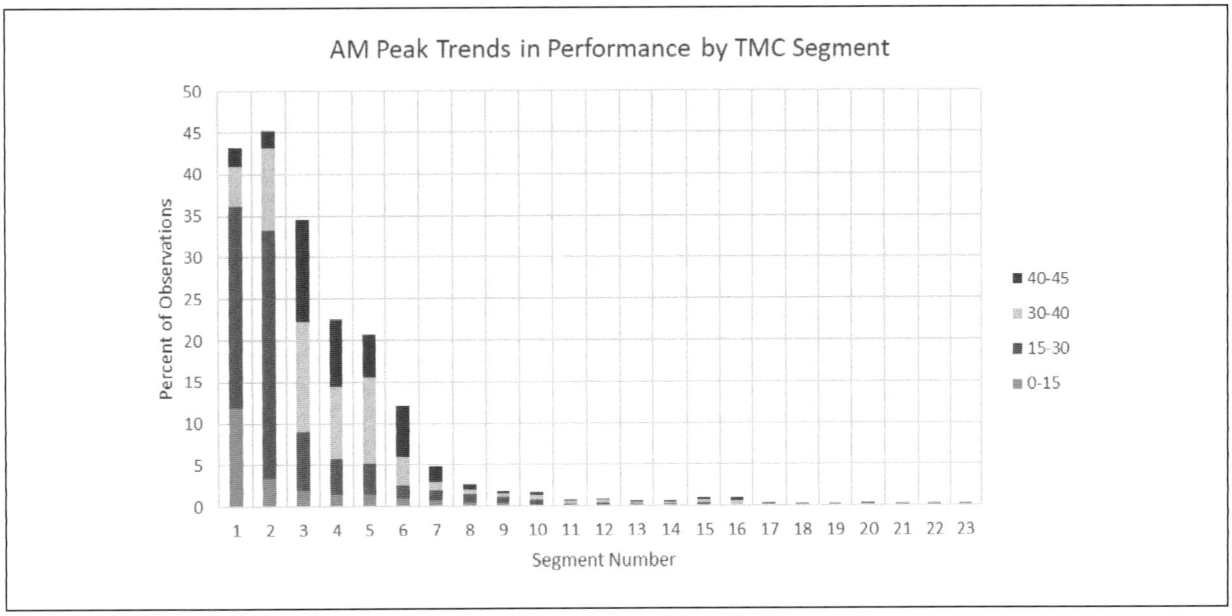

Figure 3. TMC performance in the morning peak.

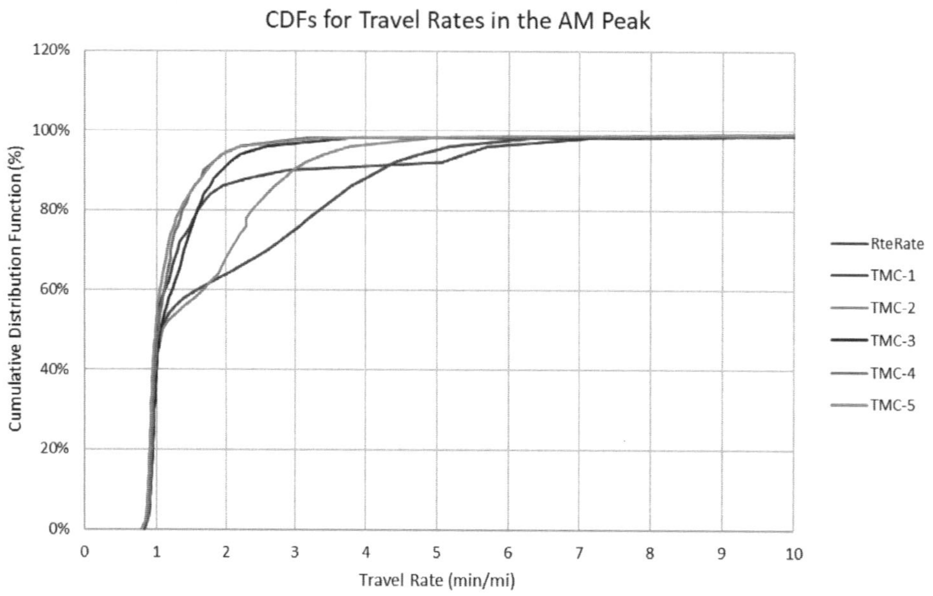

Figure 4. CDFs of travel rates for the route and the first five TMCs in the morning peak.

second TMC operates at 45 mph or less 45% of the time, at 40 mph or less 42% of the time; and 30 mph 33% of the time. This does not mean that every third day it is at 30 mph or less, but that might be the case. But it does mean that 33% of the five-minute intervals have a speed of 30 mph or less.

The CDFs of the travel rates for the route and the first five TMCs are shown in Figure 4. This again shows that the first two TMCs are the worst performers. It is also interesting to note that the CDF for the route has a plateau from about 3 min/mi to 5 min/mi. This means about 90% of the route travel rates are 3 min/mi or less but another 10% are between 5 and 6 min/mi (10–12 mph). It would be helpful if that portion of the tail could be eliminated.

The causal factors explored here are incidents and weather. Weather is defined as either freezing temperature, heavy precipitation, or both. The observations have been broken down into four operating conditions: (a) normal, (b) weather, (c) incidents, and (d) weather and incidents. Further breakdowns have been explored, for example, based on the severity of the incidents, but those results are not presented here. They do not add insight from a methodological perspective.

As can be seen from Table 3, 12,344 observations were obtained for the year. This number accounts for 98.9% of the 12,480 five-minute observations (52*5*4*12) that could have been recorded. Of the 12,344, normal operating conditions are the most prevalent. They comprise 10,363 of the total, which is 84%. tRate is the travel rate in minutes per mile. W&I is a condition that involves both weather and an incident. Weather is the next most prevalent (11.2%) followed by incidents (2.5%) and weather and incidents (2.3%). In addition, the distributions of the observations, for both the overall route and TMC-1, suggest that the weather and incident conditions, especially when the two are combined, involve higher travel rates than the normal condition (Tables 2 and 3). This is expected, but the data show the expectations are defensible.

Figures 5 and 6 show the CDFs of the travel rates for the route and TMC-1 during the morning peak on weekdays. For the route, the distributions for the normal condition and weather are almost identical. But in the case of segment 1, the weather condition distribution is worse. In both cases, the incident and incident plus weather CDFs are significantly different from those for either normal or weather, with the weather plus incident distribution being significantly different from incidents alone.

These findings demonstrate the importance of incident management protocols. Any incident mitigation programs and/or procedures deployed by the agency would be helpful. Moreover, when those incidents occur during adverse weather conditions, the impacts are more severe and actions that mitigate those impacts would have significant value.

Table 2. Distribution of Route Travel Rates by Operating Condition

Morning peak route travel rate distributions

tRate	Overall	Normal	Weather	Incident	W&I
0.0	0	0	0	0	0
0.5	0	0	0	0	0
1.0	6303	5329	874	70	30
1.5	3189	2786	305	56	42
2.0	1143	977	96	34	36
2.5	270	232	15	12	11
3.0	271	232	12	18	9
3.5	34	22	2	5	5
4.0	12	4	2	5	1
4.5	13	8	0	3	2
5.0	8	3	0	5	0
5.5	501	433	35	10	23
6.0	170	133	11	11	15
6.5	109	83	9	10	7
7.0	8	2	0	4	2
7.5	70	45	7	5	13
8.0	77	31	8	16	22
8.5	3	1	2	0	0
9.0	54	13	1	20	20
9.5	2	2	0	0	0
10.0	2	1	0	0	1
1000.0	105	26	10	24	45
Total	12344	10363	1389	308	284

Table 3. Distribution of TMC-1 Travel Rates by Operating Condition

Morning peak TMC-1 travel rate distributions

tRate	Overall	Normal	Weather	Incident	W&I
0.0	0	0	0	0	0
0.5	0	0	0	0	0
1.0	5404	4490	813	70	31
1.5	1928	1676	193	47	12
2.0	563	487	46	23	7
2.5	594	513	55	16	10
3.0	820	718	69	13	20
3.5	848	739	65	19	25
4.0	757	666	50	13	28
4.5	478	417	30	17	14
5.0	391	343	28	8	12
5.5	176	134	14	12	16
6.0	112	83	8	9	12
6.5	52	30	4	7	11
7.0	31	15	3	3	10
7.5	26	13	1	3	9
8.0	28	5	3	13	7
8.5	23	5	3	8	7
9.0	23	9	1	2	11
9.5	24	6	0	8	10
10.0	13	2	0	4	7
1000.0	53	12	3	13	25
Total	12344	10363	1389	308	284

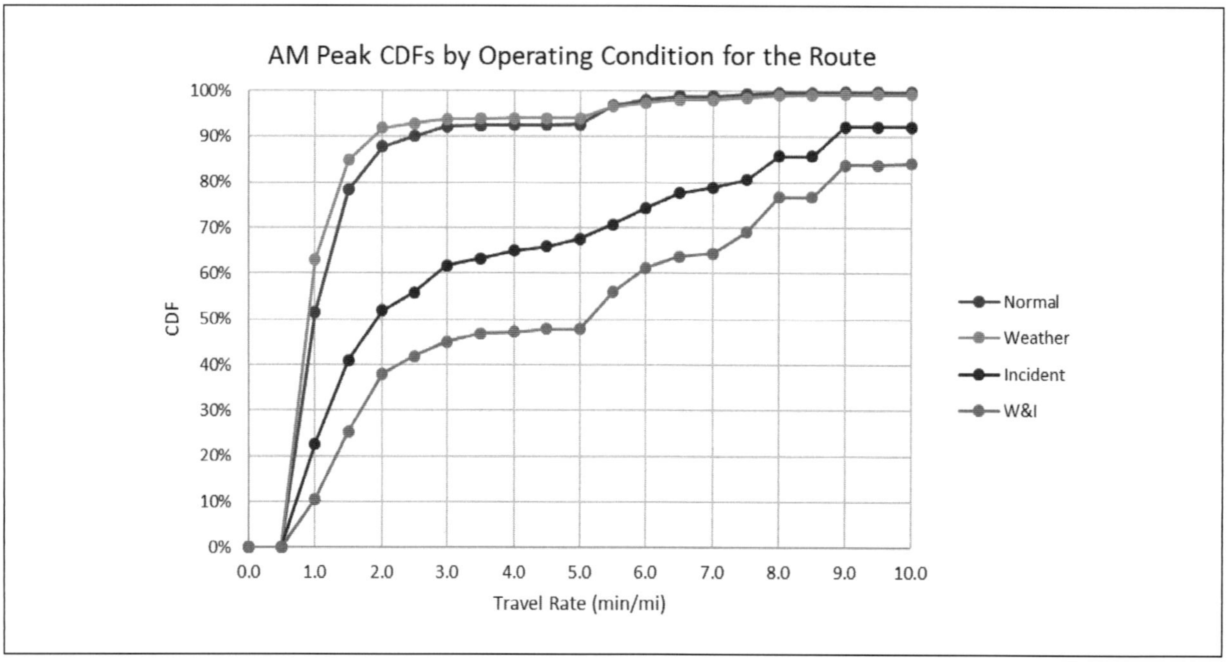

Figure 5. Morning peak CDFs by operating condition for the route.

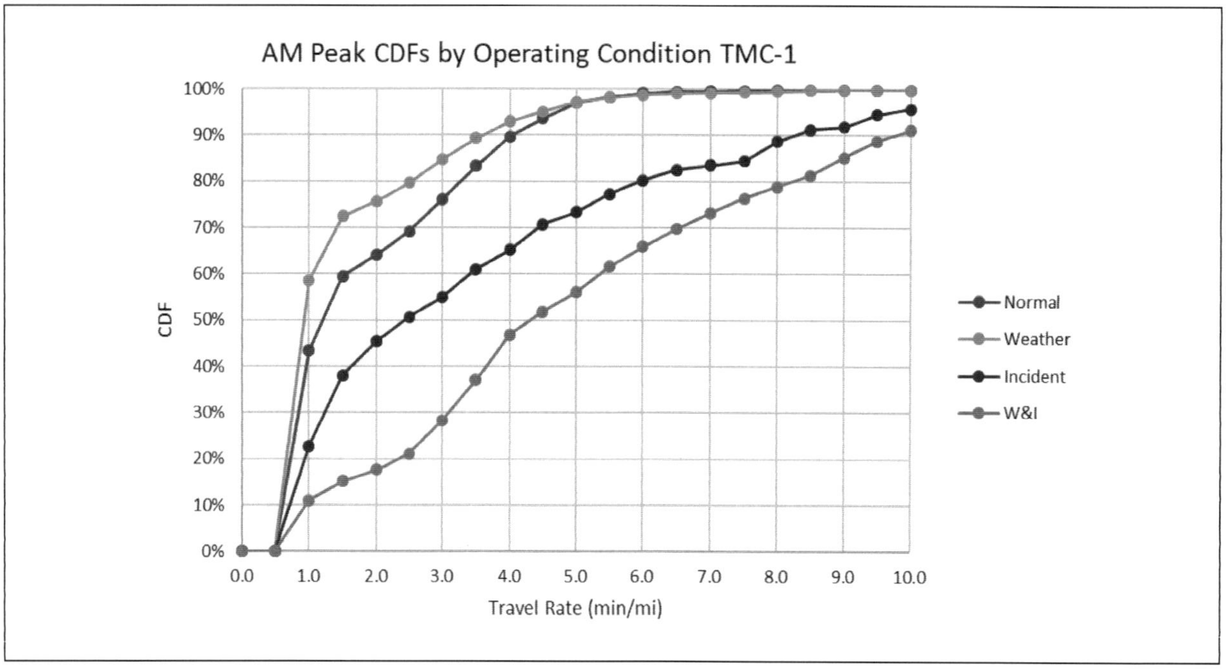

Figure 6. Morning peak CDFs by operating condition for TMC segment 1.

Conclusions and Recommendations

This paper has described the development of a prototype travel time reliability monitoring system focused an interstate highway facility in the Research Triangle region of North Carolina. It showed how the data were obtained, what data files were created, how these data were fused to create a composite database, and what insights were obtained by analyzing the data. Data for 23 TMC segments along I-40 westbound were presented to illustrate how the monitoring system can be used.

The methodology employed was the one presented in the L02 study report (*6*). The year examined was 2010. A westbound route along I-40 was selected for use in the prototype. Six interrelated datasets were developed: (i) travel rates, (ii) flow rates, (iii) incidents, (iv) weather, (v) maintenance activity, and (vi) planned events. Two primary keys were employed: time (which pertained to all six datasets; at a five-minute granularity) and TMC segment (for the first four datasets). "Stitched" travel times for both TMC segments and routes were then developed from the raw speeds (*19*). These stitched values reflected average travel times that would have been experienced by vehicles that commenced their trips across the TMC segmenst or on the overall route starting in specific five-minute intervals.

A finding from the analysis presented is that 90.9% of the time, the observed speeds are 60 mph or greater on the 23 TMCs studied. Only 8.1% of the time are they less than 60 mph. On the weekends, the performance is even better. However, during the morning peak, the speeds are lower; 81.3% of the time the speeds are 60 mph or greater, and 13% of the time (more than once every two weeks) they are 45 mph or less.

However, these statistics are somewhat misleading because no trips traverse only one TMC segment. All paths across the facility use multiple TMCs. For the route that traverses all 23 TMCs, only 79.3% of the time is the speed of this route 60 mph or greater; 13.4% of the time it is 45 mph or less. Moreover, in the morning peak, only 51.1% of the time is it 60 mph or better, and 41.3% of the time it is 45 mph or less. Hence, if the policy performance of this freeway is set to be 45 mph or better, then two out of every five weekdays that performance target would not be met. This route-based assessment is very important. It reveals problems that the TMC-based analysis misses.

CDFs for the TMCs quickly showed which ones were the source of the unreliability. For example, one of the TMCs operated at 45 mph or less 45% of the time, at 40 mph or less 42% of the time; and at 30 mph 33% of the time.

Breakdowns by causal factor revealed other trends. The causal factors explored were incidents and weather. Weather was the most prevalent cause of abnormal operating conditions (11.2%) followed by incidents (2.5%) and weather and incidents (2.3%). Insofar as impacts were concerned, for the route, the distributions of travel rate for the normal condition and weather were almost identical. But in the case of one TMC, the CDFs for the operating conditions showed that the weather condition was worse. In both the case of the overall route and this TMC, the incident and incident plus weather CDFs were significantly worse than that for either the normal or weather-affected operating condition, with the weather plus incident distribution being significantly different from incident alone.

The implications of these findings are at least three-fold. First, examining the CDFs for the travel rates helps with understanding the performance of individual facilities and with comparing performance among different facilities and for different operating conditions. Second, differentiating the CDFs among operating conditions is helpful and informative. Not only does it highlight differences in performance, it also motivates ideas about when and how mitigating actions might be taken to improve performance. Third, focusing exclusively on the performance of the TMC segments belies one's ability to gain a true sense of the quality of service to be provided by the system. All trips involve the use of multiple TMCs. And the integrated effect of the TMC performance on the travel time distribution for overall trips suggests that the trip performance is likely to be worse. Agencies would be wise to select representative routes and monitor the performance of these routes in addition to the performance of the TMCs.

In the future, this analysis could be taken further in many ways. One involves conducting multi-year assessments so that trends in performance across time can be observed, and mitigating actions identified. Another involves examining specific operating conditions in more detail so that the causal effects are better understood and more precise mitigating responses can be identified.

Acknowledgments

The results presented here have been developed from NCDOT Research Project 2016-32, SHRP2 L38 Reliability Data and Analysis Tools Implementation Assistance Program Proof of Concept Pilot Study. The research project was sponsored by FHWA and administered by North Carolina Department of Transportation (NCDOT).

Author Contributions

The authors confirm contribution to the paper as follows; study conception and design: GFL, RS; data collection: RS, GFL; analysis and interpretation of results: GFL, RS, NR, BW; draft manuscript preparation: GFL, RS. All authors reviewed the results and approved the final version of the manuscript.

References

1. Federal Highway Administration. *How Is Transportation Performance Monitoring Being Implemented?* https://www.fhwa.dot.gov/tpm/about/how.cfm. Accessed July 2017.
2. Transportation Research Board. *The Second Strategic Highway Research Program (2006–2015)*. http://www.trb.org/StrategicHighwayResearchProgram2SHRP2/Blank2.aspx. Accessed July 2017.

3. Federal Highway Administration, *Two Performance Management Final Rules Take Effect*. 2017. https://www.fhwa.dot.gov/tpm/rule.cfm. Accessed July 2017.
4. National Archives and Records Administration. *National Performance Management Measures: Assessing Performance of the National Highway System, Freight Movement on the Interstate System, and Congestion Mitigation and Air Quality Improvement Program*. https://www.federalregister.gov/documents/2017/01/18/2017-00681/national-performance-management-measures-assessing-performance-of-the-national-highway-system. Accessed July 2017.
5. Boyce, D. A Memoir of the ADVANCE Project. *Intelligent Transportation Systems Journal*, Vol. 7, No. 2, 2002, pp. 105–130.
6. *SHRP 2 Report S2-L02-RR-1: Establishing Monitoring Programs for Travel Time Reliability*. Transportation Research Board of the National Academies, Washington, D.C., 2014.
7. Barkley, T., R. Hranac, and K. Petty. Relating Travel Time Reliability and Nonrecurrent Congestion with Multistate Models. *Transportation Research Record: Journal of the Transportation Research Board*, 2012. 2278: 13–20.
8. Guo, F., Q. Li, and H. Rakha. Multistate Travel Time Reliability Models with Skewed Component Distributions. *Transportation Research Record: Journal of the Transportation Research Board*, 2012. 2315: 47–53.
9. Fraley, C., and A. E. Raftery, *MCLUST Version 3 for R: Normal Mixture Modeling and Model-Based Clustering*. Technical Report No. 504. Department of Statistics, University of Washington, Seattle, WA, 2009.
10. Transportation Research Board. *Pilot Testing of SHRP 2 Reliability Analytical Products*. 2014. http://apps.trb.org/cmsfeed/TRBNetProjectDisplay.asp?ProjectID=3457. Accessed July 2017.
11. Schrank, D., B. Eisele, T. Lomax, and J. Bak. *2015 Urban Mobility Scorecard*. Texas A&M Transportation Institute, College Station, TX, 2015.
12. Mathew, J., D. Krohn, H. Li, C. Day, and D. Bullock. *Implementation of Probe Data Performance Measures*. Report No. PA-2017-001-PU WO 001. Pennsylvania Department of Transportation, Harrisburg, PA, 2017.
13. Wolniak, M., and S. Mahapatra. Data- and Performance-Based Congestion Management Approach for Maryland Highways. *Transportation Research Record: Journal of the Transportation Research Board*, 2014. 2420: 23–32.
14. Redmill, K., and B. Coifman. System Management and Monitoring – Temporal Evaluation of Freeway Management Systems. *Proc., 2006 IEEE Intelligent Transportation Systems Conference*, Toronto, ON, Canada, IEEE, New York, NY, 2006.
15. Shi, Q., and M. Abdel-Aty. Big Data Applications in Real-Time Traffic Operation and Safety Monitoring and Improvement on Urban Expressways. *Transportation Research Part C: Emerging Technologies*, Vol. 58, 2015, pp. 380–394.
16. van Lint, J. W. C., and H. J. van Zuylen. Monitoring and Predicting Freeway Travel Time Reliability using Width and Skew of Day-to-Day Travel Time Distribution. *Transportation Research Record: Journal of the Transportation Research Board*, 2005. 1917: 54–62.
17. van Lint, J. W. C., H. J. van Zuylen, and H. Tu. Travel Time Unreliability on Freeways: Why Measures Based on Variance Tell Only Half the Story. *Transportation Research Part A: Policy and Practice*, Vol. 42, No. 1, 2008, pp. 258–277.
18. Washington State Department of Transportation. *Best Time to Leave*. http://www.wsdot.com/traffic/seattle/traveltimes/reliability/. Accessed July 2017.
19. I-95 Corridor Coalition. *Probe Data Analytics (VPP Suite)*. http://i95coalition.org/projects/probe-data-analytics/. Accessed July 2017.
20. Chase, R. T. Jr., B. Williams, N. Rouphail, and S. Kim. Comparing Reported Travel Times between Temporally-Stitched and Simultaneous Methods at Various Aggregation Intervals. Presented at 91st Annual Meeting of the Transportation Research Board, Washington, D.C., 2012.
21. I-95 Corridor Coalition. *Regional Integrated Transportation Information System (RITIS)*. http://i95coalition.org/projects/regional-integrated-transportation-information-system-ritis/. Accessed July 2017.

The Standing Committee on Freeway Operations (AHB20) peer-reviewed this paper (18-05744).

The opinions presented in the paper are those of the authors, and do not necessarily reflect the views of NCDOT.

Article

TRR
JOURNAL OF THE TRANSPORTATION RESEARCH BOARD

Transportation Research Record
2018, Vol. 2672(14) 39–49

DOI: 10.1177/0361198118784136
journals.sagepub.com/home/trr
SAGE

Methodology to Identify On-Ramps for Metering with Consideration of System-Wide Recurrent and Non-Recurrent Congestion

Homa Fartash[1], Mohammed Hadi[1], and Raj Ponnaluri[2]

Abstract

State agencies have developed warrants and guidelines for the metering of freeway on-ramps. However, these warrants only consider the traffic conditions in the vicinity of each on-ramp without considering the need to meter multiple ramps to mitigate the impacts of bottlenecks downstream. The warrants do not employ detailed analyses of traffic conditions or take advantage of the increasing availability of data from multiple sources. In addition, the existing local warrants only consider recurrent conditions with no consideration of the benefit of metering during non-recurrent events such as incidents and adverse weather. This study aims to develop a methodology for the identification of the ramps to meter that considers system-wide recurrent and non-recurrent traffic conditions based on detailed analysis of traffic data. This methodology incorporates the stochastic nature of the demand and capacity and the impacts of incidents and weather using Monte Carlo simulation and a ramp selection procedure based on a linear programming formulation. The method allows the calculation of the minimum number of ramps that need to be metered to keep flows below capacity on the freeway mainline, while keeping the on-ramp queues from spilling back to the upstream arterial street segments. The methodology can be used in conjunction with the existing local warrants to identify the ramps that need to be metered. In addition, it can be used in benefit–cost analyses of ramp metering deployments and associated decisions such as which ramps to meter and when to activate metering.

Ramp metering is a traffic management strategy that aims to prevent traffic-flow breakdown by regulating the traffic entering a freeway, using traffic signals installed at on-ramps (*1*). Preventing traffic-flow breakdown enables the freeway to operate at full capacity and thus improves traffic performance in terms of travel time, travel time reliability, safety, and environmental impacts (*2*). On the other hand, long queues on the on-ramps may spill back to the arterials as a result of ramp metering. Therefore, it is crucial to select the ramps to meter and to apply appropriate metering strategies and rates to realize the maximum benefits of ramp metering and to avoid the adverse effects of on-ramp queues (*3*). Guidelines and warrants have been developed to support the decisions as to which ramps to select for metering. These warrants consider the local conditions of each on-ramp as an isolated element, focusing on geometry and traffic conditions at the on-ramp and its merge area. Most current applications of ramp metering, however, implement advanced algorithms which consider system bottlenecks in calculating the metering rates. Such implementations lead to much better results compared with the implementation of local metering algorithms. This points to a gap between existing agency metering warrants for the installation of ramp signals and the algorithms utilized for controlling the ramp signals. In addition, the warrants do not employ the detailed analysis of traffic conditions that is now possible with the increasing availability of data from multiple sources. Moreover, the existing local warrants only consider recurrent conditions to justify ramp metering installation with no consideration of the benefits of metering during non-recurrent events such as incidents and adverse weather.

This study aims to develop a methodology for the identification of ramps to meter that considers system-

[1]Department of Civil and Environmental Engineering, Florida International University, Miami, FL
[2]Florida Department of Transportation, Tallahassee, FL

Corresponding Author:
Address correspondence to Homa Fartash: hfart001@fiu.edu

wide recurrent and non-recurrent traffic conditions based on detailed analysis of traffic, incident, and weather data. This methodology considers the stochastic nature of the demand and capacity and the impacts of incidents and weather using Monte Carlo simulation and a ramp selection procedure based on a linear programing formulation. The method allows the identification of the minimum number of ramps that need to be metered to keep flows below capacity on the freeway mainline, while keeping the on-ramp queues from spilling back to the upstream arterial street segments. The methodology can be used in conjunction with the existing local warrants to identify the ramps that need to be metered. In addition, it can be used in benefit–cost analyses of ramp metering deployments and associated decisions such as which ramps to meter and when to activate the metering.

Literature Review

This section briefly reviews the considered parameters in existing warrants for installing ramp meters followed by a review of ramp metering operation algorithms and the corresponding process for calculation of metering rate.

Existing Warrants for Installation of Ramp Metering

The metering warrants developed by state agencies in the United States consider only the traffic, geometry, and safety conditions in the immediate vicinity of the subject on-ramp (local conditions). Specifically, the traffic criteria used by different states have included measures such as mainline volume, ramp volume, the sum of the mainline and ramp volumes, mainline speed, volume/capacity ratio, level of service (LOS), occupancy, and/or density (*1–8*). Florida (*2*) and Texas (*1*) consider the mainline volume of one or two right-most lanes, in addition to the total mainline volume, since merging maneuvers have higher impacts on the lanes closer to the on-ramp. States apply different criteria and thresholds. For instance, Arizona (*4*), Florida (*2*), and Nevada (*7*) specify a more conservative speed threshold for ramp metering installation (< 50 mph for half an hour, which is around LOS D), compared with some states like Virginia, which considers speeds below 30 mph, corresponding to LOS E, to justify the installation of ramp metering. Fartash et al. (*9*) highlighted the gap between existing ramp metering warrants and the ramp metering operation algorithms which are used during operations and demonstrated the need to develop a methodology to consider system conditions, in addition to local conditions, when selecting ramps for metering. Other countries, such as Australia, the United Kingdom, and the Netherlands, utilize similar warrants for ramp metering deployment. In Australia, ramp metering is justified when the mainline volume exceeds a threshold of 1,700 vphpl. The criteria used in the United Kingdom and the Netherlands utilize the hours of congestion and the average mainline speed and volume as measures for justifying ramp metering installation (*2*).

Consideration of Bottlenecks during Operations

Ramp metering algorithms, as implemented during operations to reduce the impacts of traffic bottlenecks, can be categorized as local or system-wide algorithms. Local algorithms, such as ALINEA, percent-occupancy and demand-capacity algorithms, only target a single ramp, as an isolated element (*3, 10*). In contrast, system algorithms consider multiple ramps as a system to address the traffic congestion and breakdown at the bottleneck locations to benefit the entire system. System algorithms are proven to be more effective than local algorithms because they distribute the metering burden to several ramps, instead of relying on the metering of the immediate single on-ramp upstream of the bottleneck, which may not be sufficient and can cause long queues on the ramp. System algorithms can be grouped into cooperative, competitive, and integral algorithms. Cooperative algorithms such as Helper, linked ramp, and HERO, calculate the metering rates based on local traffic conditions followed by further adjustment based on system-wide conditions (*10–12*). HERO coordinates local traffic responsive ALINEA blocks at a system-wide level (*12*). Competitive algorithms such as FLOW, Zone, SWARM, and Seattle, calculate two independent metering rates for each ramp. One rate is based on local conditions and the other is based on system-wide conditions. The algorithm then selects the more restrictive value between the two (*3, 11, 13–15*). Integral algorithms such as METALINE and fuzzy logic incorporate the local and system-wide conditions simultaneously in the selection (*14*).

Ramp Metering and Non-Recurrent Conditions

Transportation system management and operation agencies have realized the need to activate metering during non-recurrent events like incidents and adverse weather conditions (*16*). Previous studies showed that non-recurrent conditions contribute significantly to congestion and unreliability of the transportation system.

The effects of incident on freeway capacity are well studied (*17, 18*). Goolsby (*19*) concluded that incidents with one and two lanes blocked out of three lanes reduce the capacity of the freeway mainline by 50% and 79%, respectively. Hadi et al. (*20*) investigated capacity drop for different incidents and concluded that, just after the occurrence of an incident with one lane blocked out of three lanes, the freeway capacity drops by 54 to 60%. The Highway Capacity Manual (*21*) also provides estimates of the expected reduction in capacity as a function of the number of blocked lanes (or shoulder) and the number of directional lanes. These estimates are used in this study and further discussed in the methodology section. Qi and Smith (*22*), proposed a beta distribution with an average capacity reduction due to incident of 63% and 77% with one and two out of three lanes blocked, respectively.

Haj-Salem and Papageorgiou (*23*) demonstrated that ramp meters can reduce the recurrent and non-recurrent congestion on freeway mainline. They showed that when incidents occurred, ramp metering activation could increase the mean speed by 21.5%. Bogenberger and May (*24*) showed how some dynamic metering algorithms can be updated to address the impact of an incident.

Adverse weather is also a major cause of congestion (*25*, *26*) and ramp metering activation can be used to mitigate this congestion. Ibrahim and Hall (*27*) showed that the freeway mainline speed drops by 3 to 5%, 14 to 15% and 30 to 40% for light precipitation (both rain and snow), heavy rain, and heavy snow, respectively. Stralen et al. (*28*) investigated the impacts of adverse weather on the probability of traffic-flow breakdown. They incorporated both the supply and demand aspects of the influence of adverse weather on traffic conditions using a panel mixed logit model. The average probabilities of traffic-flow breakdown for dry weather and heavy rain were estimated to be 50% and 77.4%, respectively. Li et al. (*29*) showed 6 to 12% speed reduction for freeway and arterial facilities, depending on the level of rain intensity. The HCM (*21*) also presents percentages of capacity reduction for freeway facilities under medium and heavy rain.

Case Study

The case study area is a corridor of about 15 miles including eleven on-ramps along Interstate I-95 in Broward County, Florida, from Hallandale Beach Boulevard to Commercial Boulevard. For this study, the researchers collected flow and speed data from traffic detectors and aggregated at 15-minute intervals, as well as incident data. The data was collected for the northbound direction of the corridor from the traffic management data warehouse of the Florida Department of Transportation (FDOT) for the morning and afternoon peaks (7:00–10:00 a.m. and 3:00–6:00 p.m., respectively), over a one-year period. Note that the flow data on the on- and off-ramps are collected from a planning database and only include tube count data collected for three days in a year. Unlike the mainline, continuous detector data are not available for the ramps in the traffic management data warehouse. Weather data was obtained from the National Climate Data Center.

Methodology

The method developed in this study to select ramps for metering is based on a linear programming formulation used as part of the FREQ traffic analysis tool to select ramp metering rates that ensure that demand does not exceed capacity on freeway mainline segments (*30*, *31*). The method was extended in this study to consider the stochasticity of the demand and capacity of the freeway segments, in addition to considering incident and weather impacts that result in non-recurrent congestion. The methodology starts by identifying the locations of bottlenecks. It then uses real-world detector traffic data to derive the distributions of the demand and capacity at each bottleneck location under recurrent conditions. These distributions are used as inputs to a Monte Carlo process to generate combinations of demand and capacity values at the bottleneck locations, representing the variations in the variable values from day-to-day. The values generated for each scenario are used as inputs to the linear programming formulation and the output of exercising the formulation are used to support the decisions to meter ramps under recurrent and non-recurrent conditions. The following subsections describe the methodology in more details.

Deriving Demand and Capacity Distributions and Generating Experiments of the Monte Carlo Simulation

As mentioned above, in this study, scenarios are generated using a Monte Carlo process to account for the stochastic nature of demand and capacity that results in variations in system performance between days even when there are no incidents, weather events, and/or construction activities. Monte Carlo is a popular statistical analysis method for considering the impacts of stochastic variables (*30*). To conduct the analysis, statistical distributions were fitted to historical real-world traffic data of capacity and demand measurements from the case study corridor, as explained below.

Stochastic Capacity Analysis. Many studies have addressed the stochastic nature of the freeway capacity and showed that, even under the same external conditions, various values of capacity can be observed (*32*, *33*). To consider the bottleneck stochastic capacity in this study, first each recurrent bottleneck location is identified based on data of days with no events such as incidents or bad weather, and work zones. The bottlenecks are identified by detecting low speeds being propagated upstream, with free flowing (or near free flowing) conditions occurring downstream. This identification is accomplished by examining the historical speed profiles at multiple locations upstream and downstream of the bottleneck (*34*). The main bottleneck location on the subject corridor, in both peaks, was identified as the Commercial Boulevard exit.

The next step is to estimate the capacity distribution function. This study utilizes the product limit method (PLM) by Kaplan and Meier (*35*), to estimate the distribution. The PLM is used to estimate the probability of survival past given time points (*35*). In this case, the given time point refers to traffic-flow breakdown and the lifetime variable is the capacity. Based on this method, the distribution function of the capacity could be derived as in Equation 1 (*36*):

$$F_c(q) = 1 - \prod_{i:q_i<q} \frac{k_i - d_i}{k_i}; i \in B \quad (1)$$

where;

$F_c(q)$ = Distribution function of capacity c
q = Traffic volume (vph)
q_i = Traffic volume for interval i (vph)
k_i = Number of intervals with a traffic volume of $q \geq q_i$
d_i = Number of breakdowns at a volume of q_i
B = Set of breakdown intervals

Set B includes traffic volumes at which the average speed is higher than the threshold speed of 45 mph at time interval i and in the following time interval $(i + 1)$, the average speed drops below the threshold speed. Note that if the traffic is congested at the downstream cross section during the time interval of i or $i-1$, this case is excluded from set B, because interval i does not contain any information for the capacity assessment at the observation point (*36*).

In this study, the PLM is applied to one year of archived speed and flow data. A maximum likelihood function is used to find the best-fit distribution.

Stochastic Demand Analysis for Recurrent Conditions. Traffic demand is also a stochastic variable and this stochasticity is considered in this study. To derive the stochastic distributions of freeway demands, the historical detector volume data for the same period used in the capacity analysis is used.

If the speed for the considered time interval is lower than a speed threshold (45 mph) a queue occurrence is identified at the detection station. The distance between two adjacent detectors is divided into two equal lengths and each length is assigned to the nearest detector. If the aggregated speed over 15 minutes for a detector is less than the speed threshold, then the length assigned to that detector would be added to the queue length. Therefore, the demand for each time interval is calculated using Equation 2:

$$D_i = F_i + Q_i - Q_{i-1} \quad (2)$$

where:

D_i = Demand for time interval i
F_i = Flow for time interval i
Q_i = The entire queue length for time interval i
Q_{i-1} = The entire queue length for time interval $i-1$

The set of calculated demand values are used as inputs to a MATLAB distribution fitting tool to fit the demand distribution for the morning and afternoon peak periods and derive their probability density functions. The capacity and demand distributions, derived based on archived point detector data, are presented in Figure 1 for the main bottleneck location of the case study corridor (Commercial Boulevard exit). MATLAB uses the MLE method to fit the distribution. A *p*-value is generated as

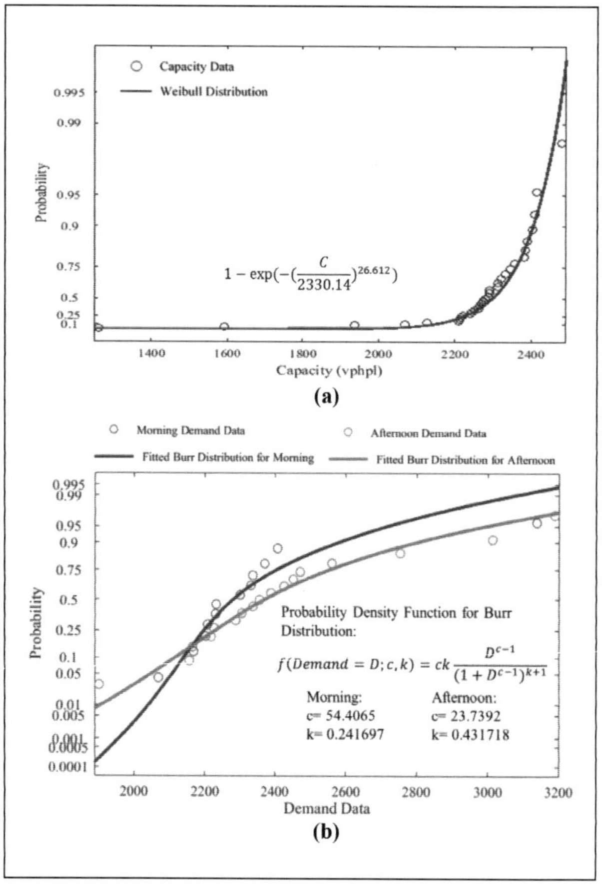

Figure 1. Capacity and demand distributions: (a) capacity distribution of the case study; (b) demand distribution for the morning and afternoon peaks.

the goodness of fit measure. If the p-value is lower than the level of significance, it can be concluded that the distribution is a good fit to the data. Various distributions were tested at the 5% level of significance and the findings for best fit are presented in Figure 1.

Monte Carlo Analysis for Recurrent Conditions. Monte Carlo analysis utilizes random sampling to conduct a large number of experiments (*37*). In each experiment, instances of the input random variables (capacity and demand, in this case) are generated based on their distributions to represent one realization of traffic conditions of the case study facility (*38*). Each realization presents a single day scenario. Then, the linear programming formulation is applied for each realization and the outputs from all realizations are further analyzed to support the decision-making process. The results from executing the linear programming formulation utilizing the outputs from each Monte Carlo experiment are the ramps that need to be metered and the metering rate for each on-ramp. Performing this process for a large number of experiments will generate a set of output variables that allow the identification of the distribution functions of these variables which can be used in the analysis.

Formulation of the Identification of Ramps for Metering

To select the ramps to be metered, all the ramps in a study area are included in the linear programming optimization. The inputs to the linear programming formulation are: the bottleneck capacity, the demands on the mainline entry point, on- and off-ramps, and the mainline exit point of the system. The optimized metering rate for each ramp in the system is the output of the model. The objective of the optimization is to maximize the vehicle throughput to keep the flow as close to capacity as possible, but lower than capacity in order to prevent traffic-flow breakdown. The objective function and constraints of the linear programming are presented in Equation 3:

$$\text{Objective Function}: MAX \sum_{i=1}^{n} x_i \tag{3}$$

Constraints:

$$\sum_{i=1}^{n} \beta_i x_i + e - \sum_{j=1}^{l} \gamma_j y_j \leq B_k$$
$$x_{i,min} \leq x_i \leq V_i$$
$$x_{i,min} = V_i - \alpha S_i$$

where:

x_i = Optimum ramp metering rate for on-ramp i (vph)

β_i = Portion of vehicles entering from on-ramp i which pass through the bottleneck location

n = Number of the on-ramps upstream of the subsection k

l = Number of off-ramps upstream of the subsection k

y_j = Ramp demand for off-ramp j (vph)

γ_j = Portion of the upstream mainline volume (e) which exit from off-ramp j

e = Upstream mainline volume (vph)

B_k = Capacity of subsection k (vph)

m = Number of subsections

V_i = Demand rate at on-ramp i (vph)

S_i = Available storage at on-ramp i (vph)

$x_{i,min}$ = Minimum metering rate for on-ramp i (vph)

α = User specified portion of the maximum queue to be accommodated by the storage length

The first constraint in Equation 3 requires that the mainline volume at a subsection k (the total of the entering flow from the on-ramps and upstream mainline minus the exiting flow from the off-ramps) be less than the capacity of this subsection. In this study, the capacity and demand values utilized in the formulation are considered as stochastic variables and are generated from their distributions using Monte Carlo simulation, described earlier. The ramp volumes are sampled in the Monte Carlo simulation based on the variability observed for the mainline demands. This is because the ramp volumes are available for only three days in a year, which does not allow obtaining the demand distribution for each ramp. If detector data is available for the on- and off-ramps for a lengthy period of time, then the distributions can be derived for each ramp and used in the Monte Carlo simulation. The linear programming formulation is assessed with the values generated from each experiment of the Monte Carlo analysis. The second constraint in the formulation requires the metering rate to be more than the minimum metering rate and less than the demand of the subject ramp. The third constraint defines the minimum metering rate for each on-ramp as equal to the difference between the demand of the on-ramp and a multiplier (defining a portion) of the available queue storage (S_i), which is the number of vehicles which can be stored in the distance from the metering stop line to a point on the ramp that is close to the upstream end of the ramp. The queue storage length in vehicles is converted to feet by assuming that each queued vehicle and the associated spacing occupy 25 feet and considering the ramp queue length and the number of lanes on the ramp. The period of the analysis is a user input and was set to one hour in the case study since, according to the analysis of the real-world data, the period of congestion in the morning and afternoon does not exceed one hour.

The linear programming formulation described in Equation 3 was solved in a manner that ensures the selection of the smallest subset of ramps for metering that meets the formulation constraints. In cases when the optimization does not have a feasible solution to eliminate traffic breakdown and congestion, the metering rates are set as the minimum feasible rate and the excessive volume above capacity is calculated. The allocation of the metering rate for the ramps starts from the closest on-ramp upstream of the bottleneck; if the minimum metering rate on this ramp is enough to prevent breakdown at the bottleneck location, other ramps need not be metered. However, if application of the minimum metering rate on this ramp is not sufficient to prevent breakdown, the next on-ramp will be considered for metering as well. The calculated rates from the linear programming are fed to an external module to modify the metering rates to ensure an equal demand-to-capacity ratio on the on-ramps. We conducted 1,000 Monte Carlo experiments to generate the capacity and demand values used as inputs to the formulation in Equation 3. The output, for each realization, are the ramps to be metered and the associated metering rates for each on-ramp. It should be noted that the ramp metering rates produced as part of the process are not expected to be used in the real-world applications in most cases since adaptive ramp metering strategies will most likely be used. However, the metering rates can be used for an initial assessment of ramp metering impacts and benefit–cost analysis described later in this paper.

Ramp Metering Selection under Non-Recurrent Conditions

The procedure described above to determine the ramps to meter under "recurrent conditions" can be extended to determine the ramps that need to be metered during incidents and weather events that produce "non-recurrent" congestion. The only difference is including the impacts of non-recurrent events (incidents and weather) in the capacity and possibly the demand inputs to the ramp selection formulation in Equation 3.

According to the HCM (*21*), once a lane is blocked due to an incident, the remaining lanes will not function at full capacity. The HCM recommended values for use as the remaining capacity of the unblocked lanes for different combinations of total number of lanes and the number of blocked lanes on freeway segments. These values are used in this study to modify the capacity distribution under normal (event-free) conditions. The impact of incidents on demand is also estimated by estimating the diversion rates during incidents using a relationship that estimates the average diversion rates during incidents for various combinations of total number of lanes and the number of blocked lanes, as developed by Hadi et al. (*39*) based on data from the study area. As an example, for an incident which results in blockage of one lane out of four lanes, the utilized relationship estimates that the demand decreases by 11.07%. The diversion percentage can be estimated using other models or modified by the analyst based on local conditions.

According to the HCM (*21*), rainfall reduces the capacity of the freeway as a function of the intensity of the rainfall in inches per hour (in/h). The HCM presents the remaining capacity of the freeway segment for the medium and heavy rain categories and these values are used in this study. The demand is assumed not to change during rain in the case study, but this change can be also considered. Table 1 summarizes the coefficients which are used to modify the demand and capacity values for each incident and rain scenario for the four-lane test segment of this study, utilizing the approach described above.

Estimation of System-Wide Ramp Metering Benefits

The benefits of the proposed methodology can be evaluated in terms of delay savings. For this purpose, queuing theory equations are used to calculate the delays for both the freeway mainline and on-ramps (*40*). The queue and associated delay on the mainline forms due to traffic-flow breakdown at the bottleneck locations. Queues due to metering are those that occur on the on-ramps due to ramp demand exceeding ramp capacity with metering. The delay saving for each scenario is calculated as the difference between the delays with and without metering. Based on a previous analysis of incident data for the case study corridor, the average incident durations utilized in the analysis are 40 minutes for shoulder incidents and 50 minutes for incidents with lane blockages (*41*). The total delay (TD) on the freeway mainline and each on-ramp is calculated using Equations 4 and 5, based on queuing theory analysis (*40*):

$$TD = t_R t_Q(\lambda - \mu_R)/2 \tag{4}$$

$$t_Q = t_R(\mu - \mu_R)/(\mu - \lambda) \tag{5}$$

where:

λ = Arrival rate (demand) (vph)

μ = Capacity rate under recurrent condition (vph)

μ_R = Reduced capacity rate under metering for ramps or under non-recurrent conditions for the mainline (vph)

t_R = Event duration (hr)

t_Q = Time in queue duration (hr)

Table 1. Capacity and Demand Adjustments for Non-Recurrent Conditions

Event	Remaining capacity for the segment	Remaining demand for the segment
Shoulder only incident	0.85 × (4 lanes) × original capacity (vphpl)	Original demand (vph)
1 lane blocked incident	0.77 × (3 lanes) × original capacity (vphpl)	(1 − 0.1107) × original demand (vph)
2 lanes blocked incident	0.50 × (2 lanes) × original capacity (vphpl)	(1 − 0.1688) × original demand (vph)
Medium rain	0.90 × (4 lanes) × original capacity (vphpl)	Original demand (vph)
Heavy rain	0.82 × (4 lanes) × original capacity (vphpl)	Original demand (vph)

The delay savings can be converted to monetary values to be incorporated into benefit–cost analyses, as described in a later section.

Results

This section presents the results from applying the discussed methodology to the I-95 corridor segment between Hallandale Beach Boulevard and Commercial Boulevard in Broward County, FL.

Demand-to-Capacity Ratio Distributions Derived Based on Monte Carlo Results

As mentioned above, 1,000 independent Monte Carlo experiments were produced with different demand and capacity values for the morning and afternoon peak periods. The distribution of the demand:capacity (D/C) ratio was generated based on the experiment results. Figure 2 shows the cumulative density functions (CDF) fitted to the D/C values for the morning and afternoon peak hours during recurrent (no-event) conditions. According to these distributions, it is estimated that the median D/C ratio is about 0.995 in the morning peak and 1.039 in the afternoon for no incident conditions. The 85th D/C ratios are 1.108 and 1.184 for the two peaks, respectively.

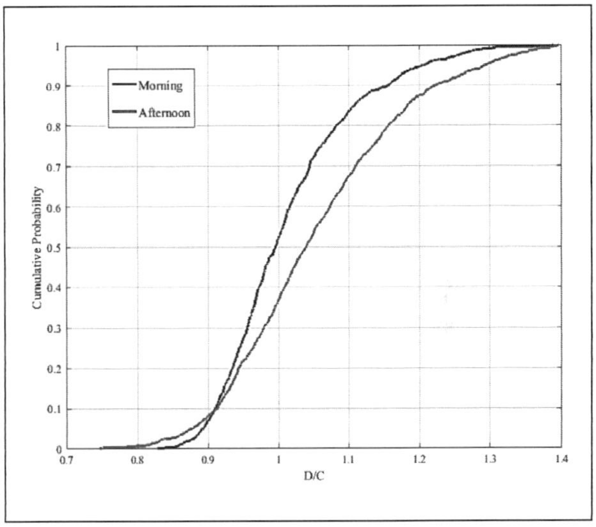

Figure 2. CDF of D/C ratios for the morning and afternoon peaks during recurrent conditions.

Determining the Number of Ramps to Meter

The capacity and demand values from the 1,000 experiments (representing days) of the Monte Carlo simulation were used as inputs to the ramp metering selection formulation for the morning and afternoon periods. The outputs from the ramp selection formulation are the number of the ramps to be metered upstream of the bottleneck and the metering rates. Statistical analysis of the total number of ramps to be metered for the morning and afternoon peaks is presented in Table 2. As shown, the results from the simulation indicate that 129 of the 1,000 experiments with no incidents or weather events will require the metering of all 11 ramps between Hallandale Boulevard and Commercial Boulevard in the morning peak and 299 will require metering at all ramps in the afternoon peak. About half of the days in the morning peak and two thirds of the days in the afternoon peak will require two or more ramps to meter. The ramp metering selection procedure was repeated with incident and weather events at the bottleneck location and the results are also presented in Table 2. The results show that with lane blockage incidents and heavy rain during the peak hours, all 11 ramps will have to be activated in most cases.

Benefit–Cost Analysis of Ramp Metering Deployment

The described methodology can be used to conduct benefit–cost analysis of the deployment of ramp metering and associated decisions such as which ramp to meter and when to activate metering. A summary of calculated delay saving of ramp metering is presented in Table 3 for recurrent and non-recurrent conditions. As mentioned above, the delays are calculated for the two peak periods. The duration of the rainfall is assumed to be one hour and the duration of the incident to be 40 minutes for shoulder incidents and 50 minutes for lane blockage incidents. Table 3 indicates that ramp metering can provide

Table 2. Total Number of Ramps to Meter for Recurrent and Non-Recurrent Conditions

Number of metered ramps	Recurrent condition		Non-recurrent (incident)						Non-recurrent (rainfall)			
			Shoulder blocked		1 lane blocked		2 lanes blocked		Medium rain		Heavy rain	
	Morning	Afternoon	Morning	Afternoon	Morning	Afternoon	Morning	Afternoon	Morning	Afternoon	Morning	Afternoon
0	523	370	0	0	0	0	0	0	0	0	0	0
1	0	0	0	0	0	0	0	0	0	0	0	0
2	3	15	19	18	0	3	0	0	1	7	0	1
3	78	40	22	10	1	4	0	0	28	19	0	2
4	63	50	47	15	1	3	0	0	61	25	0	4
5	57	25	54	7	4	4	0	0	54	19	4	7
6	30	42	46	24	3	12	0	0	66	24	8	8
7	34	43	63	27	4	2	0	0	81	53	15	17
8	18	24	43	18	17	15	0	0	53	21	11	10
9	30	18	98	12	26	7	0	0	71	18	30	12
10	35	74	68	33	46	33	0	0	96	90	54	33
11	129	299	540	827	898	908	1000	1000	437	647	878	895
Total	1000	1000	1000	1000	1000	1000	1000	1000	948	924	1000	989

significant benefits in terms of mobility and reliability, as indicated by the 50% and 85% delay savings. During recurrent conditions, the median delays decreased by 859 vehicle-hours (about 7.8 minutes/vehicle) and 1,174 vehicle-hours (about 10.3 minutes/vehicle) during the morning and afternoon peak hours, respectively. The delay savings are even greater during shoulder and one-lane blockage incidents and rain events that occur during the peak hours, as shown in Table 3. In the case of two-lane blockage incidents, ramp metering is not capable of preventing the breakdown of traffic flow, therefore, a queue forms on the mainline and reduces the delay savings.

The benefit–cost analysis of the metering is calculated by obtaining the monetary value of the delay savings as the multiplication of the median delay savings in vehicle-hours by the value of time, assumed to be $13.0 per passenger-hour, and passenger occupancy set to two passengers/vehicle based on previous studies in the region. The annual benefit of the system-wide metering in the case study area is calculated assuming 48% and 63% of the days with no events having recurrent congestion in the morning and afternoon peaks, respectively, as obtained from the Monte Carlo analysis discussed earlier; 12% and 5% of the weekdays were assumed to have additional delays due to incident and rain events, respectively. The cost values were obtained from FDOT. The project life is assumed to be 7 years and the interest rate to be 6%. Table 3 includes the parameters for benefit–cost analysis. The results show that benefit–cost ratio of the analysis is about 9.4. It should be mentioned that a similar benefit–cost analysis was conducted for system-justified versus locally justified metering and a benefit–cost ratio of 2.06 was obtained for using system-justified metering.

Conclusions

This study develops a methodology for considering system-wide conditions in the decisions to select freeway on-ramps for metering and concludes that it is crucial to consider system-wide conditions in selecting locations to install signals. In most cases, metering the ramp immediately upstream of the bottleneck is not sufficient to prevent breakdown of traffic flow on the freeway mainline. A ramp may not be a candidate for ramp signal installation according to its local traffic conditions, however, it can play a significant role in addressing the traffic congestion at a system bottleneck located miles downstream. The methodology developed in this study can be used in conjunction with existing local warrants for installing metering to bridge the gap between the installation decision and operation and management policies and strategies. The proposed

Table 3. Delay Savings Due to Ramp Metering and the Associated Benefit–Cost Analysis

Event	Time	Percentile	Total delay with metering (veh-h)	Total delay without metering (veh-h)	Total delay saving (veh-hr)	Benefit ($)
Recurrent condition	AM	50	21	880	859	13,400
		85	372	1248	876	13,666
	PM	50	166	1340	1174	18,314
		85	615	2148	1533	23,915
Medium rain	AM	50	446	1144	698	10,889
		85	622	1532	910	14,196
	PM	50	485	1604	1119	17,456
		85	602	2432	1830	28,548
Heavy rain	AM	50	512	1638	1126	17,566
		85	647	1818	1171	18,268
	PM	50	499	2311	1812	28,267
		85	587	3193	2606	40,654
Shoulder blocked incident	AM	50	1152	3033	1880	29,328
		85	1316	3168	1853	28,907
	PM	50	1093	4264	3171	49,468
		85	1298	4403	3105	48,438
One lane blocked incident	AM	50	3447	4813	1365	21,294
		85	3783	5183	1400	21,840
	PM	50	3444	6454	3010	46956
		85	3796	6841	3045	47,502
Two lanes blocked incident	AM	50	5717	5901	183	2,855
		85	6205	6355	149	2,324
	PM	50	5691	5906	215	3,354
		85	6264	6437	173	2,699
Benefit–cost analysis results						
Benefit				Annual benefit for total of 11 ramps		$6,419,910
Cost				Capital cost for 11 ramps		$3,320,000
				Annual operation and maintenance cost for 11 ramps		$121,451
Interest parameters				Interest rate		6%
				Project life		7 years
Benefit–cost ratio						9.4

ramp selection methodology considers both recurrent and non-recurrent conditions (e.g., incident and rainfall), in addition to the stochastic natures of demand and capacity.

The results of a case study analysis show that in all the cases when demand exceeds capacity, metering on-ramps adjacent to the bottleneck location is not sufficient to prevent congestion considering the constraints on the selection of the metering rates. About 13% and 30% of the Monte Carlo experiments, representing days with different demand and capacity values during recurrent conditions, requires the selection for metering of all 11 ramps in the case study corridor for the morning and afternoon peaks, respectively. The results show that with lane blockage incidents and heavy rain during the peak hours, all 11 ramps will have to be activated in most cases. The benefits of system-wide metering are calculated in terms of delay savings and can be used to support the decision to select a subset of the on-ramps for metering.

Acknowledgments

The work presented in this paper is part of a research project funded by the Florida Department of Transportation.

Author Contributions

The authors confirm contribution to the paper as follows: study conception and design: Homa Fartash, Mohammed Hadi; data collection: Homa Fartash; analysis and interpretation of results: Homa Fartash, Mohammed Hadi, Raj Ponnaluri; draft manuscript preparation: Homa Fartash. All authors reviewed the results and approved the final version of the manuscript.

References

1. Balke, K., P. Chaudhary, P. Songchitruksa, and G. Pesti. *Development of Criteria and Guidelines for Installing, Operating, and Removing TxDOT Ramp Control Signals*. Texas Department of Transportation, Austin, TX, 2009.
2. Gan, A., K. Liu, P. Alluri, and X. Zhu. *Integrated Database and Analysis System for the Evaluation of Freeway*

Corridors for Potential Ramp Signaling. Florida Department of Transportation, Tallahassee, FL, 2011.
3. Hasan, M. *Evaluation of Ramp Control Algorithms Using a Microscopic Traffic Simualtion Laboratory, MITSIM*. Massachusetts Institute of Technology, Cambridge, MA, 1999.
4. Simpson, S., and F. Yasmin. *Ramp Metering Design Guide*. Arizona Department of Transportation, Phoenix, Arizona, 2013.
5. Caltrans. *California Ramp Meter Design Manual*. California Department of Transportation, Sacramento, CA, 2000.
6. Gaisser, T., and K. DePinto. *I-70 East Ramp Metering Assessment*. Colorado Department of Transportation, Denver, CO, 2015.
7. Jacobs Engineering Group, Inc. *Managed Lanes and Ramp Metering Manual, Part 2: Implementation Plan*. Nevada Department of Transportation, Carson City, NV, 2013.
8. Wilbur Smith Associates. *Wisconsin Statewide Ramp Control Plan*. Wisconsin Department of Transportation, Madison, WI, 2006.
9. Fartash, H., M. Hadi, Y. Xiao, T. Wang, and R. Ponnaluri. Assessing the Need for System-wide Ramp Metering Installation Warrants. *Transportation Research Record: Journal of the Transportation Research Board*, 2017. 2616: 27–38.
10. Kristeleit, T. P. *Analysis and Evaluation of Ramp Metering Algorithms*. University of the Federal Armed Forces, Munich, 2014.
11. Ahn, S., R. Bertini, B. Auffray, J. Ross, and O. Eshel. Evaluating Benefits of System-wide Ramp Adaptive Ramp-Metering Strategy in Portland, Oregon. *Transportation Research Record: Journal of the Transportation Research Board*, 2007. 2012: 47–56.
12. Faulkner, L., F. Dekker, D. Gyles, I. Papamichail, and M. Papageorgiou. Evaluation of HERO Coordinated Ramp Metering Installation at the M1/M3 Freeway in Queensland, Australia. *Transportation Research Record: Journal of the Transportation Research Board*, 2014. 2470: 13–23.
13. Hourdakis, J., and P. G. Michalopoulos. Evaluation of Ramp Control Effectiveness in Two Twin Cities Freeways. Presented at 81st Annual Meeting of Transportation Research Board, Washington, D.C., 2002.
14. Elefteriadou, L., A. Kondyli, W. Brilon, L. Jacobson, F. Hall, and B. Persaud. *Proactive Ramp Management Under the Threat of Freeway-Flow Breakdown*. National Cooperative Highway Research Program, Washington, D.C., 2009.
15. Zhang, M., T. Kim, X. Nie, W. Jin, L. Chu, and W. Recker. *Evaluation of On-Ramp Control Algorithms*. California PATH Research Report UCB-ITS-PRR-2001-36, 2001.
16. Isaai, M. T., P. Najaf, M. Lavasani, and H. Nezamianpour Jahromi. Intelligent Evaluation of Transportation Management Policies for Metropolitan Areas. Presented at 92nd Annual Meeting of the Transportation Research Board, Washington, D.C., 2013.
17. Ghaffari, A., A. Tavakkoli Kashani, and S. B. Moghimidarzi. Identification of Black Spots Based on Reliability Approach. *Scientific Journal on Traffic and Transportation Research*, Vol. 25, No. 6, 2013, pp. 525–532.
18. Najaf, P., M. T. Isaai, M. Lavasani, and J. M. Thill. Influence of Incidents on Freeway Quality of Service. *Journal of Transportation Safety & Security*, Vol. 9, 2017, pp. 178–203.
19. Goolsby, M. E. Influence of Incidents on Freeway Quality of Service. Presented at 50th Annual Meeting of the Transportation Research Board, Washington, D.C., 1971.
20. Hadi, M., P. Sinha, and A. Wang. Modeling Reduction in Freeway Capacity Due to Incidents in Microscopic Simulation Models. *Transportation Research Record: Journal of the Transportation Research Board*, 2007. 1999: 62–68.
21. *Highway Capacity Manual: A Guide for Multimodal Mobility Analysis*, 6th ed. Transportation Research Board, Washington, D.C., 2016.
22. Qi, L., and B. Smith. *Characterization of Accident Capacity Reduction*. Center for Transportation Studies, University of Virginia, Charlottesville, VA, 2001.
23. Haj-Salem, H., and M. Papageorgiou. Ramp Metering Impact on Urban Corridor Traffic: Field Results. *Transportation Research Part A: Policy and Practice*, Vol. 29, No. 4, 1995, pp. 303–319.
24. Bogenberger, K., and A. D. May. *Advanced Coordinated Traffic Tesponsive Ramp Metering Strategies*. Institute of Transportation Studies, University of California UCB-ITS-PWP-99-1, 1999.
25. Ghasemzadeh, A., and M. M. Ahmed. Driver's Lane-Keeping Ability in Heavy Rain, Preliminary Investigation Using SHRP 2 Naturalistic Driving Study Data. *Transportation Research Record: Journal of the Transportation Research Board*, 2017. 2663: 99–108.
26. Ahmed, M. M., and A. Ghasemzadeh. The Impacts of Heavy Rain on Speed and Headway Behaviors: An Investigation using the SHRP 2 Naturalistic Driving Study Data. *Transportation Research Part C: Emerging Technologies*, Vol. 91, 2018, pp. 371–384.
27. Ibrahim, A. T., and F. L. Hall. Effect of Adverse Weather Conditions on Speed-Flow-Occupancy Relationships. *Transportation Research Record: Journal of the Transportation Research Board*, 1994. 1457: 184–191.
28. Stralen, V., S. C. Calvert, and E. J. Molin. The Influence of Adverse Weather Conditions on the Probability of Congestion on Dutch Motorways. Presented at 93rd Annual Meeting of the Transportation Research Board, Washington, D.C., 2014.
29. Li, Z., L. Elefteriadou, and A. Kondyli. Modeling Weather Impacts on Traffic Operations: Implementation into Florida's Travel Time Reliability Model. Presented at 93rd Annual Meeting of the Transportation Research Board, Washington, D.C., 2014.
30. Leiman, L., M. Bouaunia, and A. D. May. Integrated System of Freeway Corridor Simulation Models. *Transportation Research Record: Journal of the Transportation Research Board*, 1991. 1320: 177–189.
31. May, A. D. *Application of Simulation Models in Traffic Management Operations*. University of California, Berkeley, Calif., 1976.
32. Ponzlet, M. *Dynamik der Leistungsfähigkeiten von Autobahnen (Dynamics of Freeway Capacity)*. Schriftenreihe

des Lehrstuhls fuer Verkehrswesen der Ruhr Universitaet Bochum, No. 16. Bochum, 1996.

33. Elefteriadou, L., R. P. Roess, and W. R. McShane. Probabilistic Nature of Breakdown at Freeway Merge Junctions. *Transportation Research Record: Journal of the Transportation Research Board*, 1995. 1484: 80–89.
34. FHWA. *Traffic Congestion and Bottlenecks: Identification, Diagnosis, Solutions*. U.S. Department of Transportation, Washington, D.C. 2015.
35. Kaplan, E. L., and P. Meier. Nonparametric Estimation from Incomplete Observations. *Journal of the American Statistical Association*, Vol. 53, No. 282, 1958, pp. 457–481.
36. Okamura, H., S. Watanabe, and T. Watanabe. An Empirical Study on the Capacity of Bottlenecks on the Basic Suburban Expressway Sections in Japan. *Proc., 4th International Symposium on Highway Capacity*, Washington, D.C., 2000.
37. Mahadevan, S. Monte Carlo Simulation. In *Reliability-Based Mechanical*, Marcel Dekker, New York, 1996.
38. Lavasani, M., X. Jin, and Y. Du. Market Penetration Model for Autonomous Vehicles on the Basis of the Technology Adoption Experience. *Transportation Research Record: Journal of the Transportation Research Board*, 2016. 2597: 67–74.
39. Hadi, M., Y. Xiao, and M. L. Rojas. Estimation of Diversion Rate During Incidents on Basis of Mainline Detector Data. *Transportation Research Record: Journal of the Transportation Research Board*, 2013. 2396: 54–60.
40. May, A. D. *Traffic Flow Fundamentals*. Prentice Hall, Englewood Cliffs, NJ, 1990.
41. Hadi, M., L. Shen, C. Zhan, Y. Xiao, S. Corbin, and D. Chen. Use of Operation Data to Evaluate the Benefits and Costs of Advanced Traffic Management Components. *Transportation Research Record: Journal of the Transportation Research Board*, 2008. 2086: 48–55.

The Standing Committee on Freeway Operations (AHB20) peer-reviewed this paper (18-02138).

Article

TRR
JOURNAL OF THE TRANSPORTATION RESEARCH BOARD

Transportation Research Record
2018, Vol. 2672(14) 50–60

DOI: 10.1177/0361198118788183
journals.sagepub.com/home/trr
SAGE

Methods for the Design of Safety Service Patrol Beats: The Florida Road Ranger Case Study

Xiaotong Sun[1], Mehrdad Shahabi[1], Grady Carrick[2], Yafeng Yin[1], Sivaramakrishnan Srinivasan[3], and Nima Shirmohammadi[4]

Abstract

This paper introduces a framework for the design and deployment of safety service patrols (SSP) on freeways. Three mathematical programming models are proposed, with the targets of minimizing fleet size and total average response time and maximizing social benefits to accommodate various design objectives. Fast heuristic algorithms based on neighborhood searching techniques are presented for solving the optimization problems. In addition, a computer tool with a graphical user interface has been developed that takes roadway segments information as the input and provides real-world recommendations for deployment. The tool was tested using nearly 800 unique Florida roadway segments, totaling over 2,000 centerline freeway miles. The results were consistent with more than 100 current beats of Florida Road Ranger service patrols, and will be used for budget forecasting for the state.

Safety service patrols (SSP), known in Florida as Road Ranger service patrols (RRSP), have long proved to be an efficient incident management strategy to detect, respond, and remove incidents on the roadways. Two key decisions are central when designing the deployment strategy for SSP. The first decision involves identifying which locations or roadways warrant coverage. In this case, factors such as available funding, traffic conditions, and incidents history can affect the location decisions for operating these services. Once a freeway segment is identified and warranted for SSP deployment, the next primary decision is how to cover and patrol the target segments efficiently with the finite resources available. More specifically, the SSP program is designed to cover a roadway segment, which is usually divided into multiple non-overlapping beats, with a certain number of patrol vehicles or trucks assigned to each one. Operational hours of SSP can be flexible due to the variation of traffic situations and available funding. Programs typically prioritize peak periods, then daytime weekday periods and, if funds are available, nights and weekends.

Two recent studies surveyed the characteristics and design of SSP nationwide (*1*, *2*). The first comprehensive survey was conducted in 2008 and provided an overall quantitative description of freeway SSP deployments in the United States. The median number of vehicles owned by each SSP program was 11 (103 respondents), the median number of vehicles patrolling at peak periods was six (101 respondents); the median number of SSP program travel miles during peak travel periods was 96 miles (92 respondents), and the median number of average headway, which is calculated by dividing route miles by the number of peak vehicles on patrol, was 14.8 miles (*1*). The second study provided the survey of SSP in the State of Florida in 2016 focusing on deployment plans at state levels (*2*). Both studies show that the majority of deployments function according to simple coverage of a certain area in a given amount of time. For instance, the Florida survey shows that decisions on beat configuration were made two decades ago when Florida's statewide program began, and most decisions were made subjectively by experience (*2*). In most of the existing Florida RRSP deployments, the roadways are divided into roughly equal parts, with interchanges as the starting and ending points of each beat for patrol vehicles to

[1]Department of Civil and Environmental Engineering, University of Michigan, Ann Arbor, MI
[2]Enforcement Engineering, Inc., Saint Johns, FL
[3]Department of Civil and Coastal Engineering, University of Florida, Gainesville, FL
[4]JetBlue Airways, New York, NY

Corresponding Author:
Address correspondence to Mehrdad Shahabi: mshahabi@umich.edu

turn around. The division is based on a certain target time for vehicles to make a bi-directional circuit, better known as circulation time. For example, beat lengths in Florida's District 2 (Jacksonville) are approximately 15 miles along the centerline, so that patrolling vehicles can make a circuit in 30 minutes at a normal driving speed. Similarly, I-95 in District 4 (Fort Lauderdale) are divided into three equal parts for vehicles to circulate each beat every 30 minutes. Measures such as traffic volume, number of incidents, and a benefit–cost ratio are used as the basis for the deployment decisions. The above examples clearly show the lack of a comprehensive framework which encompasses various operations scenarios and design expectations for optimal design of these systems.

In SSP deployment, there is always a fundamental trade-off between benefits and cost. On the one hand, patrolling more miles per vehicle would decrease the number of required vehicles and lead to lower costs. More miles and fewer vehicles, however, have a negative impact on quick response. In order to incorporate various decision variables into a planning process, researchers have developed mathematical models capable of addressing the trade-off between benefits and cost to better utilize limited funding and other valuable resources.

The goal of this paper is to leverage mathematical programming techniques to assist decision makers in generating SSP deployment design decisions. In particular, three different design models are proposed and solved in order to provide recommendations for the deployment of an SSP program along the identified segments. First, a fleet size minimization model is presented, which provides the minimum total number of patrol vehicles required to meet a given target response time. The second model attempts to minimize the total average response time given that the fleet size is specified. This model will help achieve the minimum total average response time over the whole corridor. The third model solves the deployment problem for corridors without any prior SSP deployment, where the empirical data on fleet size and response time are not available. In addition, it is worth noting that these models do not specify the types of patrol vehicles, since vehicle type is closely related to incident type and geometric conditions and thus needs to be further determined at the operation stage. Finally, all models are incorporated into a computer tool which can be used by practitioners to design their program.

The next section reviews the available literature on SSP planning followed by a detailed discussion of the models and their corresponding mathematical optimization formulations. Subsequently, numerical examples conducted in Florida are presented, followed by an introduction to the integrated computer tool that was created for guidance of Florida RRSP planning deployment . Lastly, the detailed algorithms for mathematical models are attached in an appendix for readers who are interested in the additional details of the solution algorithms for the design.

Literature Review

To date, the literature on the design and deployment of SSP using mathematical modeling is limited. Since developing a patrolling strategy is a proactive form of incident management, the information about incident locations and duration is unknown to patrols, so that it is typically gleaned from historical experience. As an example, when deciding the fleet size of a new corridor, the North Carolina Department of Transportation applied a regression model which was calibrated from historical North Carolina Incident Management Assistance Patrols (IMAP) data (*3*).

Simulation models have also been widely used to deploy new programs and to make adjustments to existing systems. Wu et al. conducted sensitivity analysis in the discrete event-based simulation to determine a deployment plan that achieves the shortest incident response time (*4*). Simulation-based optimization approaches were adopted by a number of researchers since randomly occurring incidents and patrol operation can be replicated under a simulation environment. For instance, Pal used nested partition method to optimize the fleet size for a patrolling service, and load balancing algorithm to divide the patrol roadway into multiple beats such in a way that total workload variation among all the beats is minimized (*5*).

Mathematical programming techniques have also been employed for solving the SSP problem. Under the assumption that beat configuration is given, Petty developed a mathematical model with the objective of maximizing profit: delay reduction is regarded as a benefit, and the costs for the tow truck are regarded as the total operation cost (*6*). This problem is relatively easy to solve through marginal benefit analysis, while traffic delay is captured by a fundamental queuing diagram. The proposed model considers the time-varying roadway capacity, thus provides different numbers of trucks assigned to each beat in different time periods.

A more completed min-max bi-level programming model was proposed by Yin (*7*). This approach determines an optimal fleet allocation with the objective of minimizing the maximum system travel time resulting from incidents. A heuristic iterative solution algorithm was proposed to solve this non-convex model. Yin et al. (*8*) and Lou et al. (*9*) integrated beat configuration and fleet allocation problems into a single model, in both deterministic and stochastic settings. A nonlinear mixed-integer programming model was developed to minimize the overall average incident response time. Their models consider the existence of commercial tow trucks, where number of available tow trucks as well as the largest fleet size are also provided. Finally, three heuristic algorithms were proposed to solve their models. They also extended the deterministic models into a stochastic setting which allows the number of incidents and response time to be uncertain. Considering the empirical distributions of incident occurrences and their impact on travel time, multiple scenarios were considered in their proposed modeling framework with the objective of minimizing the expected response time incurred under high-consequence scenarios. Another related work was conducted by Daneshgar and Haghani, in which they proposed a joint mixed-integer model to determine beat configuration, fleet size, and vehicle allocation concurrently, aiming at minimizing incident response time (*10*).

In addition to proactive approaches for incident management, a great body of literature has been dedicated to studying reactive strategies for incident management. Dispatching tow trucks to detected incidents from centralized depots is perhaps one of the most popular strategies. There are extensive studies that concentrate on determining depot locations, fleet size, and vehicle routings through mathematical approaches. Kim et al. provided an extensive review of the related models for this approach (*11*).

The long-term goal for strategic deployment of SSP is to increase the social benefits affordably, therefore the reliable estimation of benefits is a significant factor. In urban areas, the social benefits are mainly quantified by the monetary value of savings on traffic delays, fuel consumption, and pollution emission, and secondary incidents. The widely accepted evaluation model, the freeway service patrol evaluation (FSPE) model, was first used for freeway service patrols on California freeways (*12*). In this macroscopic analysis approach, incident delays are calculated, based on queuing theory, and benefits derived from reductions in the total delay. Performance measures are recorded for time periods with and without service patrols to determine the delay savings of the program. In rural areas where the traffic volume is low, the benefits derived from delay reduction are negligible, since no severe capacity reduction is experienced. Li and Walton indicated that the major function of SSP in low-traffic areas is to assist stalled vehicles, protect stranded drivers, and provide free road assistance services (*13*). Accordingly, the major benefit was calculated as the stranded driver's savings by using the free SSP service in lieu of a commercial towing or roadway assistance company.

With the target of providing a patrolling service instead of a dispatching strategy, our three models follow mathematical programming techniques to generate static deployment plans. Our first model attempts to search for the least number of vehicles and corresponding plan that will meet service requirement such as target response time, while the second model maximizes profits using the available fleet. The third model targets maximizing social benefits when no empirical data on fleet size and response time are available. In addition, fast neighborhood search heuristic methods have been introduced and have been embedded within a computer tool to help practitioners construct the deployment plans for SSP systems.

Design Models

Patrolling strategies involve three components: beat configuration, operation scheduling, and assignment of patrol vehicles. Beat configuration establishes the geographic boundary for the patrolling unit(s). Operation scheduling determines the operation hours and is calculated according to the different traffic conditions at different time periods, and is followed by a planning guidance (*2*). Assignment of vehicles implicates the number of service vehicles or trucks that should be used for a particular geographic expanse. In this paper, the proposed design models only consider providing beat configuration and assignment of patrol vehicles when operation hours are known in advance.

For modeling purposes, the consecutive freeway corridor is mathematically assumed as a directed graph $G(N, A)$, where N is the set of nodes and A is the set of links in the graph. Each node i represents an interchange in reality where patrol vehicles can turn around. Each link (i,j) represents a freeway segment. For each link, the freeway patrol time t_{ij} and incident frequency f_{ij} are known in advance. A beat b contains one or more continuous links that a total number of $v_b(\geq 1)$ vehicles patrol. Freeway corridor G is covered by a total number of B beats and V patrol vehicles, with $V \geq B$ (Figure 1). It is also assumed that no two adjacent beats overlap, and each patrolling vehicle only serves one beat, although in practice there may be situations where short segments are covered by two beats.

Considering the different requirements from decision makers, three objective functions are specified. The first is minimizing the size of the service patrol fleet so that it uses the least social resources and understands the realities of public finance. The second is minimizing total average response time that imposes a performance metric which is measurable, yet adjustable. The third is maximizing social benefits to build a program that does the most good, irrespective of investment. Three models accomplish these three objectives, Models P1, P2, and P3 respectively.

Fleet Size Minimization

When an incident occurs, the time that it takes a responsible agency to detect the incident is the response time.

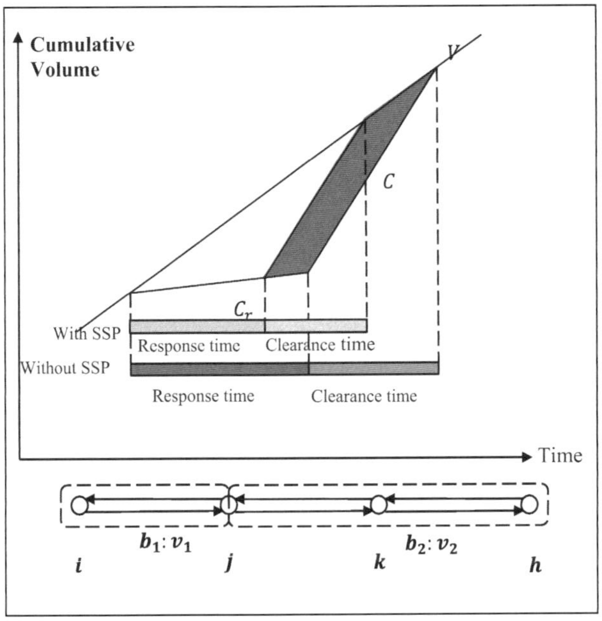

Figure 1. Fundamental queuing diagram with and without SSP and freeway corridor beat configuration.

In design of SSP systems, incident response time is an important performance measure due to its significant impact on incident duration. Under the condition in which a target response time is given, the goal of model P1 is to minimize the total fleet size of the corridor to generate the operations satisfying the given target response time.

Model P1:

$$\text{Min} \sum_b v_b \tag{1}$$

$$\text{s.t.} \sum_{ij \in A} \frac{t_{ij} x_{ij}^b}{2v_b} \leq \bar{t} \quad \forall b \in \{1, 2, \ldots, B\} \tag{2}$$

$$\sum_{ij \in A} t_{ij} x_{ij}^b \leq \hat{t} \quad \forall b \in \{1, 2, \ldots, B\} \tag{3}$$

$$\sum_b x_{ij}^b = 1 \quad \forall (i,j) \in A \tag{4}$$

$$\sum_{ij \in A} x_{ij}^b \geq 2 \quad \forall b \in \{1, 2, \ldots, B\} \tag{5}$$

$$x_{ij}^b = x_{ji}^b \quad \forall (i,j) \in A, b \in \{1, 2, \ldots, B\} \tag{6}$$

$$y_i^b \leq \sum_{j, ij \in A} x_{ij}^b \leq M y_i^b \quad \forall i \in N, b \in \{1, 2, \ldots, B\} \tag{7}$$

$$\sum_{i \in N} O_i^b y_i^b = 1 \quad \forall b \in \{1, 2, \ldots, B\} \tag{8}$$

In mixed-integer programming model (P1), Equation 2 ensures that, on each beat, the average response time does not exceed the target response time $\bar{t}$. x_{ij}^b is a binary variable, which is 1 if link (i,j) belongs to beat b and 0 otherwise. Under the assumption that the patrol vehicles in a beat are distributed uniformly in time, the average response time of an incident is assumed to be half of the headway between two successive patrol vehicles, therefore $\sum_{ij \in A} \left(t_{ij} x_{ij}^b \right) / (2v_b)$ is the average response time of beat b. Equation 3 bounds the beat length within the beat circulation time. Equations 4 to 8 specify network characteristics. In particular, Equations 4 to 6 indicate that one link can only be assigned to one beat, and the same beat must contains both directions of a road segment. Equations 7 and 8 ensure each beat has no disjointed segment, which are generated based on flow balance equations (*14*). Here, M is a sufficiently large number. Binary variable y_i^b indicates whether node i belongs to beat b, while O_i^b indicates whether node i is the starting point of beat b.

Total Average Response Time Minimization

When the total amount of budget is available, the total number of available patrol vehicles can be estimated, thus a well-designed patrolling strategy can help minimize response time to achieve the highest margins of efficiency. Model P2 configures the beat design and allocates the vehicles such that the minimum total average response time is achieved under the budget constraint.

Model P2:

$$\text{Min} \sum_b \left(\sum_{ij \in A} \frac{t_{ij} x_{ij}^b}{2v_b} \sum_{i,j \in A} f_{ij} x_{ij}^b \right) \tag{9}$$

Subject to equations 3 to 8.

$$\sum_b v_b \leq V \quad \forall b \in \{1, 2, \ldots, B\} \tag{10}$$

The notations used in P2 are consistent with those in P1. We also assume f_{ij} as the incident frequency on segment ij, $\sum_{i,j \in A} f_{ij} x_{ij}^b$ calculates the total incident frequency of beat b. Therefore, $\sum_b \left(\sum_{ij \in A} (t_{ij} x_{ij}^b / 2v_b \right) \sum_{i,j \in A} f_{ij} x_{ij}^b)$ is the total average response time to all incidents along the corridor. Thus the model objective is to minimize the total average response time. In addition, the constraints regarding the circulation time requirement and the network characteristic constraints still hold in this case. In this model, the total number of vehicles is constrained by fleet budget V.

Social Benefits Maximization

Model P3 deals with the situation in which both budget and efficiency target are unclear. In this case, neither minimizing the total average response time nor minimizing the total fleet size is feasible. Therefore, the objective of P3 is to maximize the overall social benefits.

In urban areas, the main benefits are achieved from reductions in delay, fuel consumption, and emissions. The FSPE model proposed by Sakabardonis et al. is applied to calculate such benefits (*12*). In rural or low-traffic volume areas, when an incident does not result in increase in delay, the benefits of a real-time response to an incident can be viewed as the savings in drivers' waiting time for assistance, compared with a situation in which drivers are waiting for commercial tow companies. In the formulation below, the closed-form equation of these benefits is analytically derived.

Model P3:

$$\text{Max} \sum_{b,k} B_b^{d,k} TTV_b + B_b^{f,k} F_b + B_b^{e,k} E_b^e + B_b^{s,k} S_b^k + \sum_b B_b^{w,k} WTC_b - c^* \sum_b v_b \tag{11}$$

Subject to equations 3 to 8.

The objective in Equation 11 calculates the total benefits of a determined patrol service minus the total cost. Incidents are classified into K categories by the number of lanes the incident blocked and the total number of lanes before the incident. For the kth type of incident on beat b, the social benefits include delay savings $B_b^{d,k}$, waiting time savings $B_b^{w,k}$, fuel consumption reduction $B_b^{f,k}$, and emission reduction $B_b^{e,k}$. The benefit of secondary incident reduction $B_b^{s,k}$is also counted as a type of social benefit, but it is not quantified in the calculation. TTV_b specifies the value of travel time on beat b, and WTC_b specifies the value of waiting time on beat b.

More specifically,

$$B_b^{d,k} = \begin{cases} \sum_{ij\in A} \left(\tilde{t}^2 - t_b^2\right) \frac{(V_{ij}-C_{ij}^k)(C_{ij}-C_{ij}^k)}{2(C_{ij}-V_{ij})} \left(\sum_{ij\in A} f_{ij}^k x_{ij}^b\right), if \quad V_{ij} - C_{ij}^k > 0, \\ 0, if \quad V_{ij} - C_{ij}^k \le 0. \end{cases} \quad \forall k \in \{1, 2, \ldots, K\}, b \in \{1, 2, \ldots, B\} \tag{12}$$

where $\tilde{t}$ is the average response time to an incident by commercial companies, which is applied here as a reference to the longest response time. V_{ij} is the traffic volume on link (i,j), C_{ij} is the original capacity before the incident, C_{ij}^k is the remaining capacity after the k th type of incident happens, and f_{ij}^k is the incident frequency of type k. When flow rate is greater than the remaining capacity after an incident happens, traffic delay occurs. So the benefit of the patrol service is in shortening the response time and thus reducing traffic delay. Traffic delay is calculated according to the queuing process shown in Figure 1 and the value of remaining capacity can be found in the *Highway Capacity Manual* (*15*). The reduction of traffic delay is also illustrated as the red area in Figure 1. However, if flow rate is less than the remaining capacity after an incident happens, no traffic delay will occur so that the benefit on delay savings is 0.

$B_b^{f,k} = B_b^{d,k} e_f$ is the benefit from fuel consumption reduction, where e_f is the fuel factor.

$B_b^{e,k} = B_b^{d,k} e_p$ is the benefit of emissions reduction.

TTV_b represents the value of travel time which converts benefits in a time unit into monetary value.

$B_b^{w,k} = \tilde{t} - t_b$ represents savings on waiting time, when $V_{ij} - C_{ij}^k < 0$, which means that the incident caused no traffic delay. In such a scenario, the benefits would be the time savings achieved for stranded drivers.

We should also point out that there is no unified method in the literature to quantify the benefits of avoiding secondary crashes, $B_b^{s,k}$, so that it is not incorporated in this model.

Data Collection

For the previously discussed design models (P1, P2, and P3), several parameters must be set and tuned in order to reflect real-world conditions. Incident frequency is an important term in the proposed design models, since it speaks to the important factor of when and where incidents might occur. Historical data alone cannot capture a full picture of the traffic incidents: minor incidents tend to be under-reported in places where SSP has not been deployed. A predictive model that estimates incidents from historical data given similar AADT and geometric conditions can handle this problem. With incident data from the Florida Sun Guide TMC software and the Florida Department of Transportation (FDOT) Roadway Characteristics Inventory (RCI), the number of predicted incidents for each roadway segment in Florida is possible. The incident performance functions, following the methodology of safety performance functions found in the *Highway Safety Manual*, uses negative binomial regression model (*2*):

$$N = \exp(\beta_0 + \beta_1 \times \ln(length) + \beta_2 \times \ln(AADT) + \beta_3 \times \ln(truck\ factor). + \beta_4 \times (interchange) + \beta_5 \times \ln(number\ of\ lanes))$$

The function indicates that the number of incidents is related to: segment length, number of lanes, AADT, proportion of trucks, area type, that is, urban versus rural locations, and number of interchanges. This valuable information is also provided by RCI.

The total number of incidents generated captures a broad range from minor incidents to severe crashes. For deployment purposes, the percentage of incidents assisted by SSP to the total number of incidents is calculated by studying historical data. Incident data sets are classified into four time periods: weekday daytime, weekday nighttime, weekend daytime, and weekend nighttime, corresponding to the four operation time periods. In order to capture a more accurate value of total benefits, the incident data sets are also classified by incident severity: no lane blocked (shoulder disablement), shoulder incidents, one lane blocked, two lanes blocked, and the other. The percentages vary from period to period, and from location to location. For instance, 54% of shoulder disablement in weekday daytime can be assisted, and reaching 66% if the incidents happen in weekend daytime.

In the social benefits calculation, the value of travel time can be estimated by studying the local hourly salary, percentage of trucks on roadways, and average vehicle occupancy/truck rate. Lin et al. conducted a review of Florida RRSP in 2012and provided a cost–benefit analysis listing the value of travel time in Florida per operation zone (*16*).

The disaggregate benefit for the rural motorist with a disabled car is a little more challenging. When converting waiting time into monetary value, there is no study that accurately estimates the value of the waiting time. The value of waiting time, WTC_b, is estimated based on the average price of towing or assistance services provided by commercial companies. The value of waiting time should be greater than the value of travel time.

Travel speed is used to capture the travel time of each beat. In normal conditions, the travel speed applied here is based on the posted travel speed. If the candidate corridor is located in an urban area, and the deployment is designed for weekday daytime, the average speed during peak hours needs to be adopted.

Solution Algorithm

In this work, heuristic algorithms based on neighborhood search methods are applied to solve the proposed mathematical programming formulations. A neighborhood search, also known as a local search, is a heuristic method for solving computationally hard combinatorial optimization problems. A neighborhood search starts with a feasible solution, and by applying local changes, the algorithm moves from one feasible solution to the next one in the space of candidate solutions, stopping when a certain termination criterion is reached or a time limit is elapsed. In this problem, the initial solution is generated by considering each segment as a beat. Then, by combining its adjacent segments, or links, and adjusting the number of vehicles assigned to each beat under different requirements, the algorithm will improve the initial solution until no improvement is achieved. In this case, the algorithm will converge to a local minimum; refer to the appendix for more details on the steps of the algorithms (see "Model Algorithms").

Case Study

To date, FDOT has tested the deployment analysis statewide, including all interstate highways and selected state roads. The analysis has shown that deployment plans performed can accommodate different incident severity levels. To illustrate how the models work in a real-world application, the proposed models are applied to two selected sites in Florida: I-95 in Brevard County, and I-10 from Jefferson County to Nassau County. Using two different roadways, each with multiple segments, the weekday daytime, 5.00 a.m. to 9.00 p.m. deployment was considered. Operational cost per truck was set at $56 per hour, based on recommendations by RRSP project managers. Other parameters including fuel factors, emission factors, and value of travel time used are taken from a benefit analysis report of RRSP by Lin et al. (*17*). The value of waiting time was estimated as $70 per hour, considering the average towing cost. A sensitivity analysis is conducted accordingly.

Numerical Example

Urban Areas. The first example considers a 72-mile stretch of I-95 mostly in Brevard County, which covers both urban area and rural areas around the Space Coast area of Melbourne and Cocoa Beach. This stretch is separated into 18 segments by interchanges naturally, while most segments have three lanes per direction, others have lanes lanes per direction. The average AADT is 46,462 veh/day, while the average Truck AADT is 6,611 veh/day. Urban areas usually require shorter response times compare with rural areas. In this case, we select 30 minutes as circulation time; the target response time required in Model P1 is set as 15 minutes, consistent with the estimated time applied in statewide analysis. The available fleet size required in Model P2 is set as five vehicles. Deployment plans are provided in Table 1, including starting and ending mile markers, and number of vehicles assigned to each beat. Fleet size, weighted average response time, and yearly total cost are listed for comparison. Different from average response time applied in model design, weighted average response time is the average response time of the whole study area considered varying incident frequencies among different segments.

Table 1. Deployment Schedule for I-95 Brevard County

Fleet size minimization					Total average response time minimization					Social benefits maximization				
b	l_b	v_b	Start mile mark	End mile mark	b	l_b	v_b	Start mile mark	End mile mark	b	l_b	v_b	Start mile mark	End mile mark
1	1-2	1	160	176	1	1-2	1	160	176	1	1	1	160	173
2	3-5	1	176	188	2	3-5	1	176	188	2	2-5	2	173	188
3	6-10	1	188	205	3	6-9	1	188	202	3	6-8	1	188	201
4	11-14	1	205	220	4	10-13	1	202	215	4	9-12	1	201	212
5	15-17	1	220	232	5	14-17	1	215	232	5	13-14	1	212	220
										6	15-17	1	220	232
Total fleet size				5	Total fleet size				5	Total fleet size				6
Weighted average response time (mins)				12.47	Weighted average response time (mins)				12.08	Weighted average response time (mins)				8.65
Truck cost per year ($ million)				1,092	Truck cost per year ($ million)				1,092	Truck cost per year ($ million)				1,528.8

Table 2. Deployment Schedule for I-10 Jefferson County to Nassau County

Fleet size minimization					Total average response time minimization					Social benefits maximization				
b	l_b	v_b	Start mile mark	End mile mark	b	l_b	v_b	Start mile mark	End mile mark	b	l_b	v_b	Start mile mark	End mile mark
1	1–3	1	217	241	1	1–4	2	217	262	1	1–2	1	217	233
2	4–8	1	241	275	2	5–9	1	262	296	2	3–4	1	233	251
3	9–14	1	275	303	3	10–15	1	296	327	3	5–9	2	251	283
4	15–20	1	303	336	4	16–20	1	327	336	4	10–13	1	283	301
										5	14–16	1	301	324
										6	17–20	1	324	336
Total fleet size				4	Total fleet size				4	Total fleet size				7
Weighted average response time (mins)				25.91	Weighted average response time (mins)				25.67	Weighted average response time (mins)				14.53
Truck cost per year ($ million)				873.6	Truck cost per year ($ million)				1,092	Truck cost per year ($ million)				1,528.8

Comparing the results of the three models, it is shown that all three models have provided the outcomes according to their objectives. In particular, Model P1 provides the smallest fleet size, five vehicles, to cover the entire area under the required target. This result is tested by Model P2 as well, which shows that a fleet size of four cannot cover the entire area when circulation time is 30 minutes. Furthermore, Model P2 achieves smaller weighted average response time with five vehicles assigned to five beats. The difference between Model P1 and Model P2 is beat configuration. Model P3 requires seven vehicles assigned to six beats and reaches weighted average response time of 8.65 minutes. Although this plan would lead to a higher cost, it achieves the highest social benefits according to the results.

Rural Areas. The second example is a 123-mile long stretch with 20 segments on I-10, from Jefferson County to Nassau County. This part of I-10 freeway mainly crosses rural area, where the number of incidents as well as available labor are relatively small: the average AADT is 24,671 veh/day, and the average truck AADT is 5,865 veh/day. Thus, we consider 60 minutes as the desired circulation time, and the target response time 30 minutes for Model P1.The available fleet size is four vehicles for Model P2.

Table 2 presents all three deployment plans. Model P1 shows that to meet the target response time and circulation time requirement, at least four vehicles are needed. This result is validated by Model P2 which does not provide a plan when total fleet size is three. Model P2 then gives a relatively shorter average response time compared withthat of Model P1, when fleet size is still four. Model P3 generates a plan with six beats and seven7 vehicles in which weighted average response time decreases drastically. Note that all these plans are local optimal solutions

Table 3. Sensitivity Analysis of I-95

		Fleet size (weighted average response time)			
		Circulation time			
Model P1		30	40	50	60
Target response time	15	5(12.47)	6(11.03)	6(11.03)	5(12.65)
	20	5(12.47)	4(17.51)	4(17.51)	4(17.28)
	25	5(12.47)	4(17.51)	3(20.57)	3(21.76)
	30	5(12.47)	4(17.51)	3(20.57)	3(24.20)
Model P2		Weighted average response time			
		Circulation time			
		30	40	50	60
Fleet size	3	-	-	20.5	20.5
	4	-	15.63	16.69	15.69
	5	12.08	13.39	12.08	12.17
	6	10.73	10.78	10.25	10.35
	7	9.8	12.35	8.57	9.42
Model P3		Fleet size (weighted average response time)			
		Circulation time			
		30	40	50	60
Reference response time	30	7(8.79)	6(12.63)	6(10.13)	6(10.13)

Table 4. Sensitivity Analysis of I-10

		Fleet size and weighted average response time			
		Circulation time			
Model P1		30	40	50	60
Target response time	15	9(11.67)	9(11.67)	9(11.53)	7(14.56)
	20	9(11.67)	7(15.23)	7(15.07)	7(14.56)
	25	9(11.67)	7(15.23)	6(17.72)	6(17.72)
	30	9(11.67)	7(15.23)	6(17.72)	4(25.91)
Model P2		Weighted average response time			
		Circulation time			
		30	40	50	60
Fleet size	4	-	-	-	25.67
	5	-	-	20.38	21.82
	6	-	-	18.22	17.57
	7	-	15.23	17.51	18.96
Model P3		Fleet size and weighted average response time			
		Circulation time			
		30	40	50	60
Reference response time	30	10(10.42)	8(13.81)	7(14.94)	7(14.52)

generated from heuristic algorithms which are good enough to satisfy customers' requirements; better deployment plans of beats and arrangement of vehicles may also occur under the same input conditions.

Impact of Fleet Size and Response Time

In this section the effects of changes in fleet size and response time have been investigated on the results of each of the models for the two case studies of I-95 and I-10. More specifically, in Tables 3 and 4, a sensitivity analysis has been conducted for Model P1 according to different values of circulation time and the target response time. Similarly, the sensitivity of Model P2 with respect to different levels of circulation time and fleet size has been depicted. Finally, Model P3 has been analyzed with changes in circulation time.

The results show that the shorter the target response time, the more patrol vehicles are required to respond to incidents which would consequently lead to shorter

Model Algorithms

Algorithm for Model P1

```
Begin, generate an initial solution (b, l, r, v )
  Set a binary variable flag equal to true
  While flag = true
        Add boundary links to adjacent beats and update beat configuration to (b′, l′, r′)
        Update fleet size in the greedy search manner to v′
      If the circulation time requirement is satisfied, then
          Update the beat configuration by scenario 1 or scenario 2
          (b, l, r, v) ← (b′, l′, r′, v′)
          Flag = false
     End if
  End While
  End
```

Algorithm for Model P2

```
Begin, generate an initial solution (b, l, r, v )
  While fleet size ≠fleet budget
          Generate a feasible solution (b′, l′, r′, v′)
  End While
  Update beat design by adding boundary links to adjacent beat b, the new beat configuration to (b′, l′, r′)
  If scenario 1 is met
  Fleet size of beat b equals to v_b + v_{b+1} or v_b + v_{b−1}, the number of beats decreases by 1, the new fleet allocation is v′
  If the circulation time requirement is satisfied, then
          Update the beat configuration (b, l, r, v) ← (b′, l′, r′, v′)
     End if
  End if
  If scenario 2 is met
  The fleet size of each beat remains unchanged
  If the circulation time requirement is satisfied, then
          Update the beat configuration (b, l, r, v) ← (b′, l′, r′, v)
     End if
  End if
  Improve the feasible solution by allocating patrol vehicles
  End
```

Algorithm for Model P3

```
Begin, generate an initial solution (b, l, r, v )
  Set a binary variable flag equal to true
  While flag = true
        Add boundary links to adjacent beats and update beat configuration to (b′, l′, r′)
        Update the fleet size by depth-first search to v′
      If the circulation time requirement is satisfied, then
          Update the beat configuration by scenario 1 or scenario 2
          (b, l, r, v) ← (b′, l′, r′, v′)
          Flag = false
     End if
  End While
  End
```

circulation time. We can also conclude that when the required maximum circulation time is close to target response time, then the deployment plan, specifically beat configuration, is restricted by maximum circulation time; while when target response time is far less than maximum allowed circulation time, the value of target response time dominates the final deployment plan. In addition, given a certain maximum circulation time, adding one more vehicle usually helps decrease average response time, but the margin of benefits also decreases when there are too many extra vehicles, compared with the minimum number of vehicles required to finish tasks. Compared with Models P1 and P2, Model P3 generally proposes deployment plans with larger fleet size and shorter response time. This result is reasonable since a large proportion of social benefits comes from travel time and waiting time savings associated with shorter response time. However, these deployment plans also require higher cost, which may not be realistically achievable. In practice, decision makers can always use Model P3 to

Figure 2. Design window and report window.

generate a base deployment plan which determines the largest fleet size to operate SSP, and subsequently use that value as a reference for Models P1 and P2.

User Interface

For practical purposes, an Excel tool named RRSP Integrated Interface is designed to provide planning guidelines and patrol beat design for managers responsible for contracting the RRSP operations (Figure 2). The model requires input data which is readily available from current practices of incident reporting and decision variables such as operation time periods, circulation time, target response time, or available fleet size. Deployment decisions can be generated automatically, both in an Excel worksheet and as a report which can be printed in PDF format.

Conclusion

This paper proposes three mathematical models for RRSP deployment, including beat configuration and vehicle assignment. For each model, one specific heuristic algorithm is proposed and customized. All three models are tested for various freeway segments in Florida. Heuristic methods based on neighborhood search method have been developed which are capable of handling the considered segments with their associated demand in certain time periods. In addition, an Excel-based computer tool was also developed, which can easily adapt to different requirements and changes in input parameters. Although this model framework is created for freeway segments, it can also be easily extended to a general traffic network. Future work could be unfolded from two aspects. Since currently implemented algorithms can only find a local optimal deployment plan, one potential area of research is to develop new algorithms that could find deployment plans that are closer to global optimum within a reasonable amount of time. Another possible direction is to consider the effects on normal operations when some vehicles are assisting incidents. In that case, the number of vehicles that are patrolling decreases and level of service is consequently lower. Mathematical models can be reconstructed to capture these dynamic changes during the operations.

Appendix

In the algorithms, we consider the freeway as a horizontal bi-directional linear structure consisting of multiple consecutive links. Each link s is assigned to only one beat b_s. Each beat is identified by its left link ID, l_b, and right link ID, r_b, which determines the starting and ending points together.

Acknowledgments

The work described in this paper was fully supported by Florida Department of Transportation.

Author Contributions

The authors confirm contribution to the paper as follows: study conception and design: Xiaotong Sun, Mehrdad Shahabi, Nima Shirmohammadi, Yafeng Yin; data collection: Grady Carrick, Sivaramakrishnan Srinivasan; analysis and interpretation of results: Xiaotong Sun, Mehrdad Shahabi, Nima Shirmohammadi; draft manuscript preparation: Xiaotong Sun. All authors reviewed the results and approved the final version of the manuscript.

References

1. Baird, M. E. *Overview of Freeway Service Patrols in the United States: Final Report*. Vanderbilt University, Nashville, Tenn., 2008. www.operationsacademy.org/wp-content/uploads/2016/04/FSP_FinalReportNov2008-24-Jul-2009.pdf. Accessed June 17, 2017.
2. Carrick, G., K. Jermprapai, S. Srinivasan, and Y. Yin. Development of Guidance for Safety Service Patrol Deployment Decisions in Florida. *Transportation Research Record: Journal of the Transportation Research Board*, 2017. 2660: 48–57.
3. Khattak, A. J., and N. Rouphail. *Incident Management Assistance Patrols: Assessment of Investment Benefits and Costs*. No. FHWA/NC/2005-02, 2005. https://connect.ncdot.gov/projects/planning/RNAProjDocs/2003-06FinalReportPart1.pdf. Accessed March 5, 2016.
4. Wu, W., L. Shen, X. J. Ji, and W. Jin. Analysis of Freeway Service Patrol with Discrete Event-Based Simulation. *Simulation Modelling Practice and Theory*, Vol. 47, 2014, pp. 141–151.
5. Pal, R., and K. C. Sinha. *Optimal Design of Freeway Incident Response Systems*. Publication FHWA/IN/JTRP-99/10. Joint Transportation Research Program, Indiana Department of Transportation and Purdue University, West Lafayette, Ind., 2000.
6. Petty, K. F. *Incidents on the Freeway: Detection and Management*. Doctoral dissertation. University of California, Berkeley, 1997.
7. Yin, Y. Optimal Fleet Allocation of Freeway Service Patrols. *Networks and Spatial Economics*, Vol. 6, No. 3, 2006, pp. 221–234.
8. Lou, Y., Y. Yin, and S. Lawphongpanich. Freeway Service Patrol Deployment Planning for Incident Management and Congestion Mitigation. *Transportation Research Part C: Emerging Technologies*, Vol. 19, No. 2, 2011, pp. 283–295.
9. Yin, Y. A Scenario-based Model for Fleet Allocation of Freeway Service Patrols. *Networks and Spatial Economics*, Vol. 8, No. 4, 2008, pp. 407–417.
10. Daneshgar, F., and A. Haghani. Joint Mixed Integer Model to Minimize Incident Response Time in Freeway Service Patrol Program. Presented at 95th Annual Meeting of the Transportation Research Board, Washington, D.C., 2016.
11. Kim, H., W. Kim, G. L. Chang, and S. Rochon. Design of Emergency Response System to Minimize Incident Impacts: Case Study for Maryland District 7 Network. *Transportation Research Record: Journal of the Transportation Research Board*, 2014. 2470: 65–77.
12. Skabardonis, A., K. Petty, P. Varaiya, and R. Bertini. *Evaluation of the Freeway Service Patrol (FSP) in Los Angeles*. California Partners for Advanced Transit and Highways (PATH), 1998. http://www.path.berkeley.edu/sites/default/files/publications/PRR-98-31.pdf. Accessed February 3, 2016.
13. Li, P., and J. R. Walton. Evaluating Freeway Service Patrols in Low-Traffic Areas Using Discrete-Event Simulation. *Journal of Transportation Engineering*, Vol. 139, No. 11, 2013, pp. 1095–1104. https://doi.org/10.1061/(ASCE)TE.1943-5436.0000595#sthash.gLfKUwQV.dpuf.
14. Ahuja, R. K., T. L. Magnanti, and J. B. Orlin. *Network Flows: Theory, Algorithms, and Applications*. Prentice-Hall Inc, Upper Saddle River, N.J., 1993.
15. *Highway Capacity Manual 2010*. Transportation Research Board, Washington, D.C., 2010.
16. Lin, P. S., A. Fabregas, and H. Zhou. *Review and Update of Road Ranger Cost Benefit Analysis*. Florida Department of Transportation Research Final Report BDK 84 977-15, Tampa, Fla., 2012. http://www.dot.state.fl.us/research-center/Completed_Proj/Summary_TE/FDOT_BDK84_977-15_rpt.pdf. Accessed April 15, 2016.
17. Lin, P. S., A. Fabregas, and H. Chen. Cost Benefit Analysis of Freeway Service Patrol Programs: A Case Study in Florida. *Sustainable Transportation Systems: Plan, Design, Build, Manage, and Maintain. Presented at Ninth Asia Pacific Transportation Development Conference*, Chongqing, China, 2012, pp. 1–8. https://doi.org/10.1061/9780784412299.0001

The Standing Committee on Freeway Operations (AHB20) peer-reviewed this paper (18-01074).

The opinions, findings, and conclusions expressed in this publication are those of the authors and not necessarily those of the Florida Department of Transportation or the U.S. Department of Transportation.

Article

TRR
JOURNAL OF THE TRANSPORTATION RESEARCH BOARD

Transportation Research Record
2018, Vol. 2672(14) 61–72

DOI: 10.1177/0361198118792119
journals.sagepub.com/home/trr
SAGE

Knowledge-Based System for Estimating Incident Clearance Duration for Maryland I-95

Minsu Won[1], **Hyeonmi Kim**[1], **and Gang-Len Chang**[1]

Abstract

For incident response operations to be appreciated by the general public, it is essential that responsible highway agencies are capable of providing the estimated clearance duration of a detected incident at a level sufficiently reliable for motorists to make proper decisions such as selecting a detour route. Depending on the estimated clearance duration, the incident response center can then implement proper strategies to interact with motorists, ranging from providing incident information only to executing mandatory detouring operations. This study presents a knowledge-based system, based on detailed incident reports collected by the Maryland-CHART (Coordinated Highway Action Response Team) program between years 2012 and 2016, for such needs. The proposed system features the use of interval-based estimates derived from knowledge of historical data, with different confidence levels for each estimated incident clearance duration, and its rule-based structure for convenient updates with new data and available expertise from field operators. As some key variables associated with incident duration often only become available as the clearance operations progress, the developed system with its sequential nature allows users to dynamically revise the estimated duration when additional data have been reported. The preliminary evaluation results have shown the promise of the developed system which, with its invaluable historical information, can circumvent the many data quality and availability issues which have long plagued the applicability of some state-of-the-art models on this subject.

To contend with non-recurrent congestion, caused mainly by various types of incidents, most responsible highway agencies over the past decade have implemented reasonably efficient response plans, ranging from providing information about the incident location to motorists to deploying full-scale detouring operations. Such practices, however, are not sufficient in the intelligent transportation systems era because most en-route drivers would like to have more information to make a proper route choice decision. Control center operators, aside from devoting full efforts to clear incidents, also need to decide the most cost-effective management strategy, such as offering travel time information for alternate routes for either advisory or mandatory detour operations. The difficulty in advancing the state of such practices to include these desired functions mainly lies in the lack of real-time estimated clearance duration (or entire incident duration) for detected incidents, which consequently prevents the responsible agency from estimating resulting queue lengths, impacts on the ramp flows, surrounding local arterials, and travel times.

Over the past several decades, transportation researchers have devoted considerable efforts on estimating incident duration with various techniques, including continuous statistical models (*1–16*), discrete/classification methods (*17–22*), neural network approaches (*23–30*), and hybrid modeling techniques which integrate different methods (*31–37*). Also, some have developed time-sequential models to account for the fact that incident information at the operations center is acquired over the life of the incident (*4*, *7*, *24*, *35*). The accuracy of those models is highly dependent on available variables, collected by the field response agencies, and the data quality, as well as the statistical properties of available data samples. Therefore, depending on the underlying assumptions of such models and their consistency with the distribution of collected incident data, such models mostly suffer from a lack of transferability and difficulty of use and update by the highway agencies. Thus, despite their contributions to advancing the state of the art on this subject, providing estimated incident duration during non-recurrent congestion or guiding responsive traffic

[1]Department of Civil and Environmental Engineering, University of Maryland, College Park, MD

Corresponding Author:
Address correspondence to Minsu Won: mswon@umd.edu

operations remains to be achieved in most highway networks.

Furthermore, in reviewing the incident response practices in most state highway agencies (*38*), one may recognize that a variety of factors and their complex interrelations have limited the field applications of various existing models or tools for estimating incident duration. Examples of such challenges include: (1) many variables (e.g., incident type, lane blockages, available resources) may have significant impacts on the clearance duration of a detected incident; (2) many critical factors, such as incident type (i.e., collision with fatality, collision with property damage, or collision with personal injury), responders (i.e., need for tow units, medical units, and fireboard units), and environmental conditions, are qualitative in nature and often are neither available nor collectible during the incident response and clearance process; (3) available models or tools may not be sufficiently robust to contend with the inevitable missing data or data errors caused by human factor-related issues; (4) the quantitative output from some elegant advanced models may not directly correspond to the available management actions during incident response operations; and (5) most commonly and critically, many responsible highway agencies do not have an effective system to collect various critical data associated with incident response operations in real time at the desirable level of accuracy.

In response to the above concerns and in view of the uniquely rich incident response data available in the Maryland Coordinated Highway Action Response Team (CHART) program over the past 20 years (*39*), this research has developed a knowledge-based system for robustly estimating incident clearance duration, allowing control center staff to select the most effective advanced traveler information system (ATIS) and response strategy when tackling non-recurrent congestion. The selection of the knowledge-based models is based on the following concerns: (1) most quantitative variables associated with incident duration are collected by the responsible agencies via approximation, not by precise measurement; (2) because of the emergency nature of incident response and clearance, many vital data are qualitative and/or not collected; (3) the experience and expertise of some senior responders could be very valuable in estimating the required incident duration, especially when facing a large amount of missing data; (4) the distribution of key variables in the incident database is often inconsistent with the underlying assumptions of many available statistical models; and (5) the employed method for such needs ought to be transparent and convenient for field operators to understand, evaluate, and update with most recent available information.

The proposed model, designed for field application, has the following distinct features: (1) it uses the information and knowledge embedded in a large body of earlier incident records to derive effective estimation rules which are convenient for update, and also flexible for incorporating with the experience/judgment of field operators or respondents; (2) it is constructed with a knowledge-based structure that allows efficient enhancement with additional incident cases and rules; and (3) it has an interval-based output to correspond to available traffic operations and information distribution strategies during the incident response process. A graphical interface of the proposed system is shown in Figure 1. Development procedures for all estimation rules are detailed in the following sections.

This study defines the incident clearance time (CT) as the time elapsed from arrival of the first response unit at the scene to the complete clearance of the incident. Note that the selection of clearance duration, instead of the entire incident duration, is because most incident response times under the Maryland-CHART highway patrolling strategy are quite stable, varying only in a very indifferent range.

Study Scope: I-95 in Maryland

To develop an incident CT estimation model, the research first applied the incident dataset of years 2012–2016 from the CHART II Database for derivation of estimation rules. The data of the first 4 years (2012–2015) were used for model calibration, and the final year (2016) was used for model evaluation. The data of the first 6 months (January–June) in 2017 were used for model update. As a prototype system, its spatial scope covers only MD I-95 from exit 27 to exit 109, and the data include incidents with collision only. Similar models for incidents resulting in only shoulder-lane blockage have also been developed in the final system, but not reported hereafter.

Table 1 shows the variables recorded during each incident event, which are incident type, location, time, lane blockage information, involved vehicle information, environmental conditions, response unit information, and so forth.

Data Preprocessing

To remove potential data errors by system operators, which often occur during real-time incident response and management process, the following steps were taken after consulting with field operators.

Step (1) If the difference between the event-cleared time and the all-blocked-lane-reopened time, including shoulders, is larger than 5 minutes (see Figure 2*a*), the all-blocked-lane-reopened time is taken as the event-cleared time in computing the incident CT.

Table 1. List of Key Factors Associated with the Response/Clearance of a Detected Incident

Category	Variables	Classification
Incident type	Incident type	Collision with Fatality (CF), Collision with Personal Injury (CPI), Collision with Property Damage (CPD)
Time	Hour indicator	AM-peak (7 a.m.–10 a.m.), Day time (10 a.m.–4 p.m.), PM-peak (4 p.m.–7 p.m.), Night time (7 p.m.–7 a.m.)
	Weekend indicator	Weekend, Weekday
	Holiday indicator	Holiday, Non-Holiday
	Season indicator	Spring, Summer, Fall, Winter
Location	County indicator	Prince George, Howard, Baltimore, Baltimore City, Harford, Cecil
	Direction indicator	Northbound, Southbound
	Exit number indicator	Exit 27, ..., Exit 109
Environmental condition	Pavement condition indicator	Dry, Wet, Snow/Ice, Chemical wet, Unspecified
	Hazard material related	Yes, No
Operation center	Center indicator	AOC, TOC3, TOC4, TOC5, SOC, Others
Lane blockage	# of blocked lanes*	1, 2, 3, 4...
	# of blocked shoulders	0, 1, 2, 3...
	# of blocked travel lanes**	0, 1, 2, 3...
	# of blocked traffic lanes	0, 1, 2, 3...
	# of blocked auxiliary lanes***	0, 1, 2, 3...
	Travel lane blocked in tunnel	Yes, No
	Travel lane blocked in toll	Yes, No
Involved vehicle	# of total involved vehicles****	1, 2, 3, 4...
	# of involved passenger cars	0, 1, 2, 3...
	# of involved trucks	0, 1, 2, 3...
	# of involved motorcycles	0, 1, 2, 3...
Responder	# of total response units	1, 2, 3, 4...
	# of arrived CHART	0, 1, 2, 3...
	# of arrived Police	0, 1, 2, 3...
	# of arrived Fireboard	0, 1, 2, 3...
	# of arrived Medical service	0, 1, 2, 3...
	# of arrived TOW service	0, 1, 2, 3...
	FIRST responder	CHART, Police, Fireboard, Medical, Tow

*Lanes = Shoulders + Travel lanes
**Travel lanes = Traffic lanes + Auxiliary lanes
***Auxiliary lane includes on-ramp, off-ramp, acceleration lane, deceleration lane, and collector/distributor lane
****Vehicle includes passenger car, truck, bus, cyclist, pedestrian, and motorcycle

Step (2) If the difference between the all-blocked-lane-reopened time and the all-blocked-travel-lane-reopened time is larger than 32 minutes, the sum of the all-travel-lane-reopened time plus 32 minutes is used as the event-cleared time in computing the incident CT. The logic behind this step can be seen from Figure 2*b*, in which 90% of the open times of all blocked lanes, including shoulders, for all recorded incidents are within 32 minutes difference from the open time of blocked travel lanes.

Step (3) As some recorded incident events show an unreasonably short CT for various reasons, the 10% quantile is used as the lower bound to trim the available data.

The final dataset after the data preprocessing is shown in Figure 3.

Model Development

In view of the large number of qualitative factors, the incident data were first categorized based on those major factors, and then classification rules were sought to divide the incident data into several categories based on incident CT, using the association rule mining method and hybrid modeling method, which are well fit to the knowledge-based modeling. Association rule mining method is commonly used to analyze data for frequent if/then patterns and to identify the most important relationships (*40*, *41*). It has the advantages for users to interpret the resulting relationships and to implement the results in practice. The hybrid modeling method is mainly selected to integrate different methods based on incident duration nature (*38*). It enables the developed system to provide the estimate in a sequential manner, and to deal with an incident with

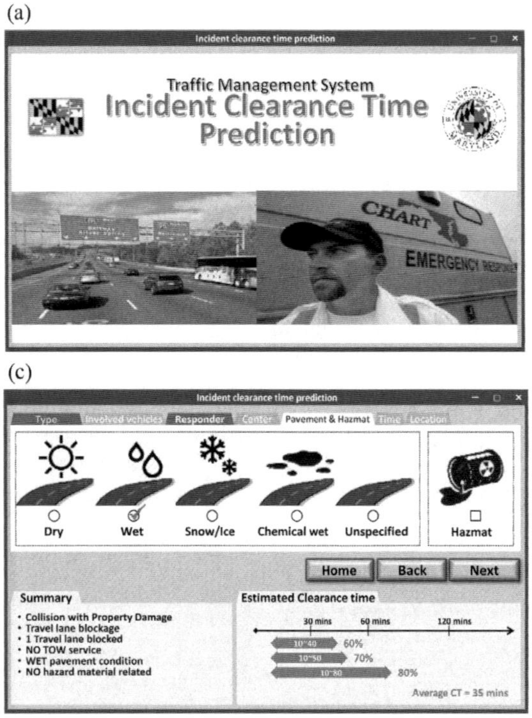

Figure 1. Examples of the software interface. (*a*) Main cover (intro); (*b*) Incident type input and updated clearance time (CT); (*c*) Pavement and Hazmat related input and updated CT; (*d*) Location input and updated CT.

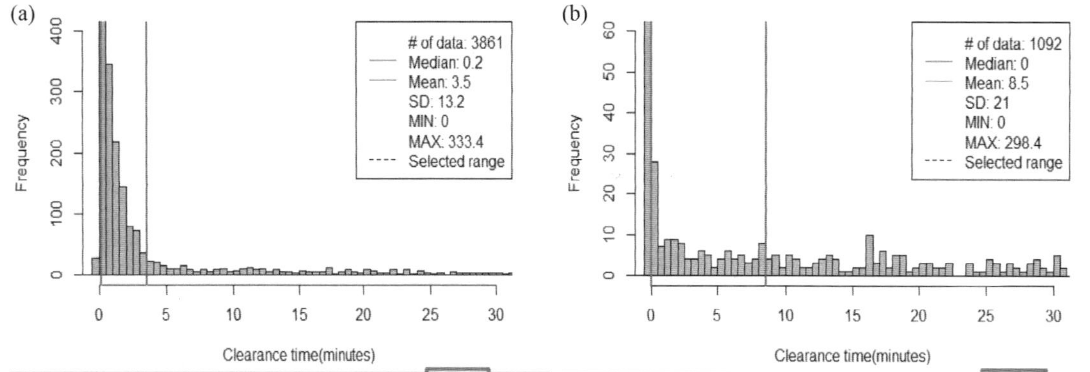

Quantile	10%	30%	50%	70%	80%	90%	95%
Time(mins)	0.05	0.08	0.17	0.83	1.63	6.77	20.53

- Incident type: Collision
- Data period: January of Year 2012 ~ December of Year 2015

Quantile	10%	30%	50%	70%	80%	90%	95%
Time(mins)	0.00	0.00	0.00	0.67	12.13	32.44	48.07

- Incident type: Collision
- Data period: January of Year 2012 ~ December of Year 2015

Figure 2. (*a*) Distribution of time differences between the all-lane-reopen times and the event-cleared times and (*b*) distribution of time differences between the travel-lane-reopen times and the all-lane-reopen times.

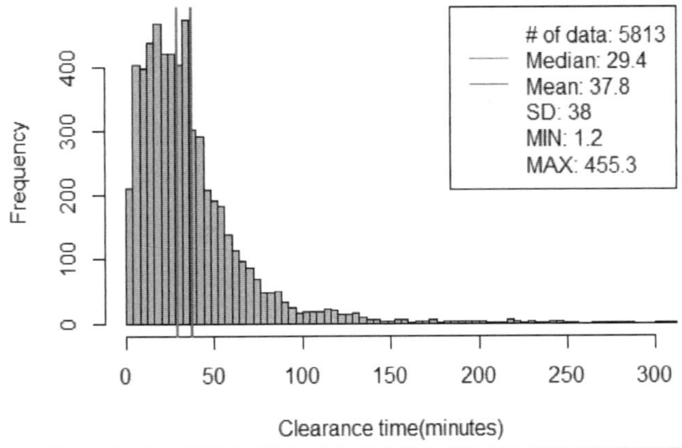

Quantile	10%	30%	50%	70%	80%	90%
Time(mins)	7.7	18.4	29.4	41.9	52.5	71.2

- Incident type: Collision
- Data period: Year 2012 ~ Year 2015

CT (mins)	Sample size		
	Original	After step 1-2	After step 3
< 30	3,475 (54%)	3,609 (56%)	2,959 (51%)
30~60	1,943 (30%)	1,990 (31%)	1,990 (34%)
60~120	831 (13%)	682 (11%)	682 (12%)
≥ 120	214 (3%)	182 (3%)	182 (3%)
Total	6,463	6,463	5,813
Mean	*36.8 mins*	*34.1 mins*	*37.8 mins*

Figure 3. Distribution of clearance times after the data preprocessing.

missing data. Moreover, from each category of incident CT, several ranges of estimated clearance duration were selected according to approximately 60%, 70%, and 80% of probability for field operators to make a proper assessment. Note that the selection of confidence thresholds can be specified by the users based on the available incident response resources and the quality of available data recorded during the incident clearance process.

Incident Categorization

Note that key factors associated with the resulting incident CT have complex relationships with one another. To sort out such relationships, all available incident data are categorized with the following steps: (1) dividing all incident data into two categories: shoulder-only blockage and travel-lane blockage; (2) providing the mean incident CT and its estimated ranges with approximately 60%, 70%, and 80% of probability; and (3) further dividing the travel-lane-blockage cases into several groups, based on the incident types and number of blocked travel lanes. Note that no further classification for shoulder-only-blockage cases was taken, because they all lie in a very stable and small range. By the same token, the incident data for Collision with Fatality (CF), resulting in the travel-lane blockage, were not further classified with the number of blocked lanes because of the small sample size.

Such an initial estimate for each detected incident (see Figure 4) may be useful for operators to assess potential impacts when the information for the full picture of the detected incident is not yet available in the clearance process.

Development of Classification Rules

Based on the incident categorization in Figure 4, the initial CT information for each incident case can be provided. In the subsequent steps, both the association mining method and hybrid modeling method are applied to the last seven categories in the travel-lane-blockage cases to increase the estimation accuracy. Note that the clearance duration for most shoulder-only blockage cases is around the interval of 30 min, so only the travel-lane-blockage cases are considered in exercising these modeling steps for the further classification.

Collision with Personal Injury (CPI) and Collision with Property Damage (CPD) in the Travel-Lane-Blockage Cases. There are six categories of CPI and CPD in the travel-lane-blockage cases, which are divided by the number of blocked travel lanes. To estimate the range of incident CT for each category for operational needs, first the data are classified into "$<$30 minutes" and "$\geq$30 minutes" using the association rule mining method. Then, from the "$\geq$30 minutes" subset, other classification rules are searched to classify the data into "$<$60 minutes" and "$\geq$60 minutes". Similarly, classification rules can be found to further categorize the expected incident CTs for "$<$120 minutes" and "$\geq$120 minutes." Finally, based on the distributions of these categories, the three intervals of

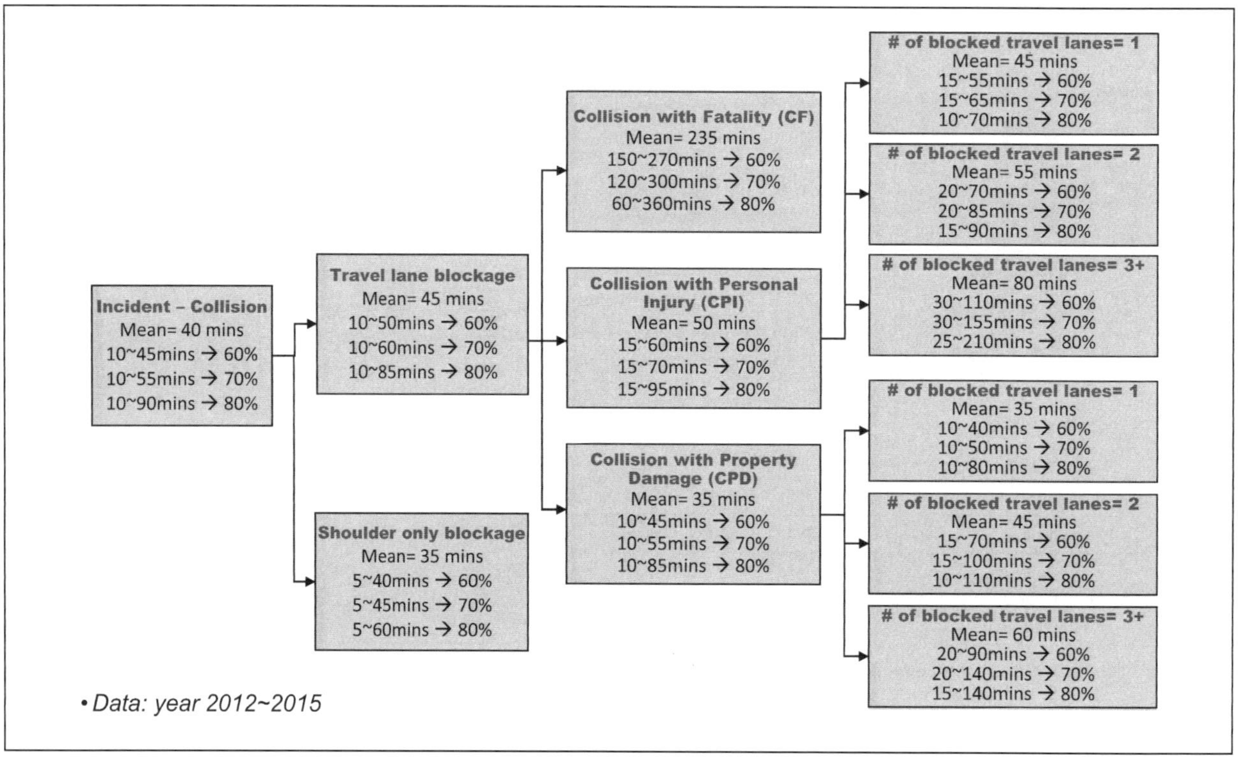

Figure 4. Initial incident categorization and estimated clearance duration.

expected incident CT under the 60%, 70%, and 80% confidence levels are produced.

The procedures to model the CPI and CPD in the travel-lane-blockage events to generate classification rules for "< 30 minutes" and "≥30 minutes" are as follows:

Step (1) From the dataset, search for classification rules to classify the data into "< 30 minutes" and "≥30 minutes" using the association rule mining method.

Step (2) Select a critical rule with approximately more than 75% confidence level and the highest support level.

Step (3) Filter out the data associated with the selected rule from the dataset and proceed with further classification for the remaining data.

Step (4) Stop the classification procedures if no further rule for classification can be derived from the remaining data.

Step (5) Otherwise, go to Step (1) and repeat the same procedures.

Figure 5 shows an example of CPI with two blocked travel lanes. Using such hybrid rules for classification, four classification rules are found for "< 30 minutes" and "≥30 minutes" and the estimated interval for incident CT is set up for the remaining data.

In the next step, the same classification procedures were applied to the dataset classified in the subset of "≥30 minutes" in the previous step to further divide the data into "< 60 minutes" and "≥60 minutes". The same procedures were also employed for "< 120 minutes" and "≥120 minutes" categories. Figure 6 shows the results of this hybrid classification procedure for six intervals of estimated incident CT, such as "< 30 minutes," "≥30 minutes," "< 60 minutes," "≥60 minutes," "< 120 minutes," and "≥120 minutes," as an example of application for CPI with two-travel-lane blockage.

Figure 6*c* shows the rule application process for the CPI with two blocked travel lanes. If an incident with CPI and two blocked travel lanes was reported, this event will go to the rules in the first table in Figure 6*c*. Based on the rules in the first table, this event will be estimated to belong to "CPI-2Lane-1" or "CPI-2Lane-2." If this event is confirmed from the field report to belong to "CPI-2Lane-1," the model will then provide the expected incident CT information associated with "CPI-2Lane-1." If not, this event will go to the rules in the second table for further classification. This is the process to provide the expected incident CT information sequentially using the proposed system. Note that this

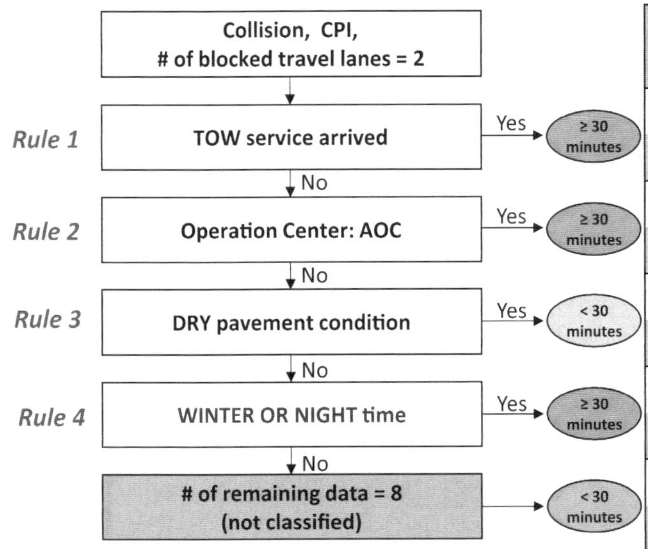

• Training set: year 2012~2015, Test set: year 2016

Set	Accuracy*	Cumulative Accuracy	Cumulative Coverage
Training	83%(83/100)	83%	46%
Test	80%(24/30)	80%	53%
Training	74%(57/77)	79%	82%
Test	80%(12/15)	80%	79%
Training	79%(22/28)	79%	94%
Test	60%(6/10)	76%	96%
Training	100%(4/4)	79%	96%
Test	-%(0/0)	76%	96%
Training	75%(6/8)	79%	100%
Test	100%(2/2)	77%	100%

Figure 5. An example of the classification rules for "<30 mins" and "≥30 mins" in the CPI with two blocked travel lanes.
*Numbers in parenthesis represent "the number of data that are correctly estimated by the rule in the remaining dataset/the number of data that is categorized by the rule in the remaining dataset."

hybrid classification model is to be executed in a sequential process, so it can provide an initial estimate in real time, even if full information for a detected incident from the field is not available.

Collision with Fatality (CF) in the Travel-Lane-Blockage Cases. There are only 25 and 5 events in the training set (2012–2015) and test set (2016), respectively, for this type of incident. Thus, the modeling procedure for the CPI and CPD in the travel-lane-blockage cases is not suitable for these CF cases. In addition, most CF cases have incident CTs longer than 120 min, so different boundaries were set for the CT ranges for such incidents. Rule-generation procedures used for analyzing CF incidents are summarized below:

Step (1) Calculate the median value of the incident CT of the dataset and divide the dataset into two groups using the median value.
Step (2) Search the classification rules to classify the data into those two groups using the association rule mining method.
Step (3) Select a critical rule with approximately more than the 70% confidence level and the highest support level.
Step (4) Divide the data into two subsets according to the selected rules and set up the estimated incident CT ranges, based on the distribution of incidents in each subset.
Step (5) Stop the procedures if the estimated incident CT range is sufficiently robust for use by control center operators.
Step (6) Otherwise, go to the first step and repeat the procedures for incident data in each subset.

This classification mining method can sequentially classify the dataset into two groups, based on the median value. Thus, it can continuously reduce the distribution and range of the incident CTs from each subset. Figure 7 shows the classification results and the estimated intervals for incident CT. Most estimated ranges for CT for collision-fatality incidents have an interval of 60 min with approximately 70% of confidence.

Model Evaluation and Updates

Model Evaluation

In this study, data over four years (2012–2015) were used as the training set for model calibration. Based on the developed model results, the model was then evaluated using the data in year 2016 as the test set. Table 2 shows the model calibration and validation results. The resulting accuracies for the estimated travel-lane-blockage incidents in the training set and test set are approximately 80% and 75%, respectively.

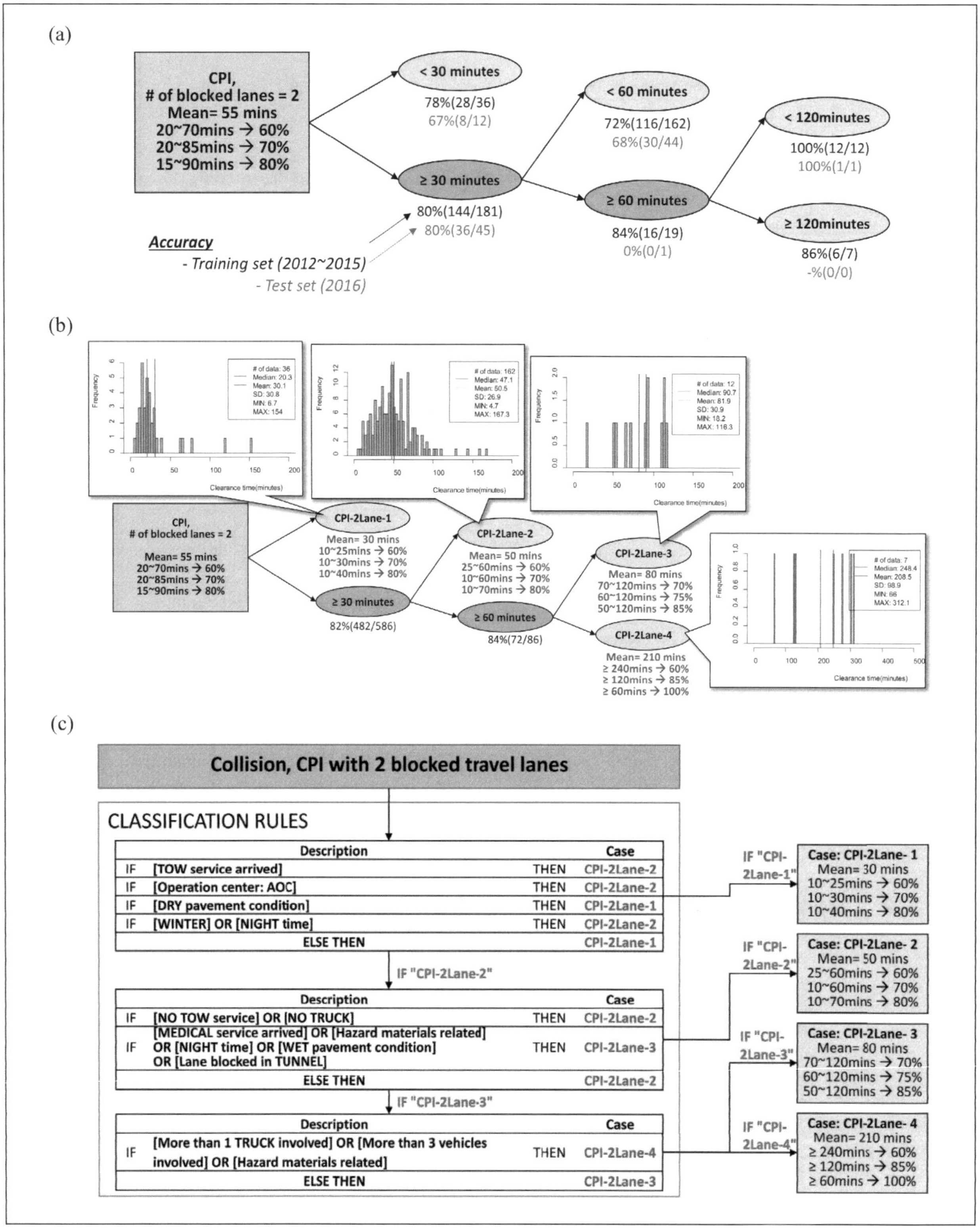

Figure 6. An example of application for CPI with two-travel-lane blockage. (*a*) Six expected incident clearance time ranges; (*b*) Distributions and ranges of incident clearance time with different confidence levels; (*c*) Classification rules and provided information.

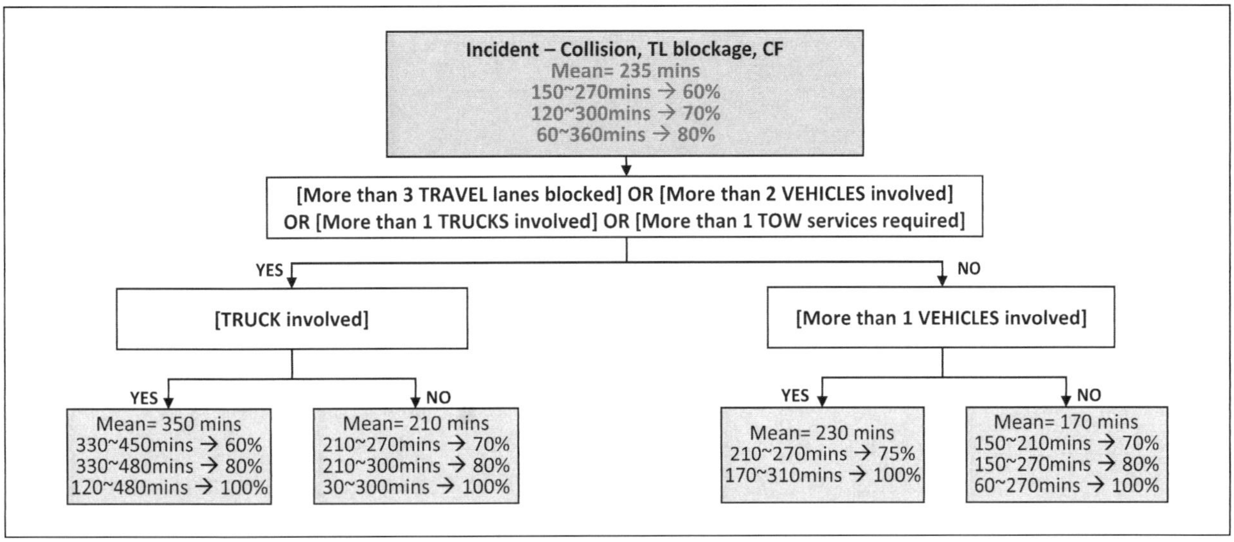

Figure 7. Classification rules and incident clearance time ranges in the CF with travel-lane blockage.

Table 2. Model Validation Results

	Travel lane (TL) blockage cases							
		CPI			CPD			
	CF	1 TL block.	2	3 +	1	2	3 +	Total
Training set (2012–2015)	100% (21/21)	79.6% (399/501)	82% (178/217)	85.9% (79/92)	77.2% (700/907)	77.8% (140/180)	80% (32/40)	79.1% (1549/1958)
Test set (2016)	83.3% (5/6)	77.5% (131/169)	71.9% (41/57)	62.5% (15/24)	74.5% (222/298)	70.3% (45/64)	75% (6/8)	74.3% (465/626)

Note: Numbers in parenthesis represent "the number of data whose clearance time is correctly estimated by the model/the total number of data in the category."

Model Updates

As new incident data are recorded daily to the database, to enhance the model's performance, the developed model was updated with the new dataset from the first 6 months (January–June) in 2017.

The model update process is focused on those misestimated cases with more than 5 min of difference between the actual and the estimated incident CTs. First, a rule was sought to discriminate those misestimated cases. Then, any newly derived rule was applied as a supplemental rule or to revise the initial model. The difference between adding the supplemental rule and revising the initial model is that a newly derived rule will serve as a supplemental rule if it covers a small portion of cases, or the estimated incident CT by this selected rule is notably different from the incident CT estimated with the initial model rules. Otherwise, the initial model will be updated with newly derived rules. In addition, to prevent the model from overfit in the update process, the impacts of the newly updated rules on the early dataset were assessed. Table 3 shows the estimation results for the six sequential model updates.

In brief, the developed knowledge-based model was able to achieve more than 85% accuracy after being updated with new data from six additional months. For example, the model accuracy for February 2017 (77.5%) increased to 85.0% by the model updated with January 2017 data. The model accuracy for March 2017 (72.7%) increased to 75.8% by the model after updated with the data of February 2017.

System Applications

The proposed system needs to have the following two unique features to account for the uncertainty associated with the nature of incident response and clearance: (1)

Table 3. Model Update Results

Model	Training set 2012–2015	Test set 2016	Additional datasets for model updates Jan. 2017	Feb. 2017	Mar. 2017	Apr. 2017	May 2017	Jun. 2017
Before updates	79.1% (1549/1958)	74.3% (465/626)	75.8% (25/33)	77.5% (31/40)	72.7% (24/33)	67.7% (21/31)	78.3% (36/46)	79.3% (23/29)
After update with Jan. 2017	79.2% (1550/1958)	74.4% (466/626)	**87.9% (29/33)**	85.0% (34/40)	72.7% (24/33)	67.7% (21/31)	78.3% (36/46)	79.3% (23/29)
After update with Feb. 2017	79.8% (1562/1958)	75.7% (474/626)	87.9% (29/33)	**95.0% (38/40)**	75.8% (25/33)	67.7% (21/31)	78.3% (36/46)	79.3% (23/29)
After update with Mar. 2017	79.8% (1563/1958)	76.0% (476/626)	87.9% (29/33)	95.0% (38/40)	**100% (33/33)**	67.7% (21/31)	78.3% (36/46)	79.3% (23/29)
After update with Apr. 2017	80.0% (1566/1958)	77.0% (482/626)	87.9% (29/33)	95% (38/40)	100% (33/33)	**87.1% (27/31)**	78.3% (36/46)	79.3% (23/29)
After update with May 2017	80.1% (1569/1958)	77.2% (483/626)	87.9% (29/33)	95% (38/40)	100% (33/33)	87.1% (27/31)	**93.5% (43/46)**	79.3% (23/29)
After update with Jun. 2017	80.2% (1570/1958)	77.2% (483/626)	87.9% (29/33)	95% (38/40)	100% (33/33)	87.1% (27/31)	93.5% (43/46)	**93.1% (27/29)**

Note: Numbers in parenthesis represent "the number of data whose clearance time is correctly estimated by the model/the total number of data."

using intervals with different confidences for estimating the clearance duration, and (2) using three thresholds of 30 min, 60 min, and 120 min to proceed with the classification of available incident data. Both key features are actually developed in response to the needs of field operators, and to be consistent with their selection of potential response plans. For example, responsible staff in the traffic control/response center are likely to provide only the incident nature and location to the general public if the clearance duration is estimated to be less than 30 min. However, more information about the detected incident and resulting traffic conditions, such as travel time information on available alternate routes, would be expected if the estimated clearance duration extends to between 30 and 60 min. By the same token, responsible traffic agencies may have to exercise advisory detouring plans and share all related information to the general public if the clearance duration of the detected incident is expected to prolong to between 60 to 120 min. In general, mandatory detour operations may be implemented for severe incidents (e.g., involving fatalities or hazardous material) if it is likely to take more than two hours to recover the traffic conditions.

Conclusion

This study has presented a knowledge-based system for estimating the interval of a detected incident's CT, using the incident data from Maryland-CHART between Years 2012 and 2015 as the calibration set and Year 2016 as the evaluation set. The proposed system features the use of interval-based estimates, derived from knowledge of historical incident response data with different confidence levels for incident CT, and the rule-based structure for convenient updates with new data and for effective incorporation of expertise from field operators to revise the initial estimate. Its sequential nature allows users to dynamically revise the estimate when additional data have been reported, because often some key variables affecting the duration of a detected incident only become available as the clearance operations progress.

The preliminary evaluation results have shown the promise of the proposed system which, with its invaluable historical knowledge, can circumvent many of the data quality and availability issues that plague the applicability of some state-of-the-art models on this subject. The proposed system can also be used to estimate reliable travel time during incidents, given the approximated or observed queue length. Furthermore, the proposed interval-based estimates with three thresholds (i.e., 30, 60, and 120 min) offer information consistent with the potential ATIS/ATMS strategies (e.g., providing incident information only, advisory or mandatory detouring operations) for the responsible highway agencies to interact with en-route motorists and the general public. Further research along this line includes: (1) extending the knowledge-based rules and embedded models to neighboring highway systems to evaluate its transferability, and (2) developing a reliable and convenient update mechanism for experienced field operators to enhance and enrich the knowledge base.

Acknowledgments

This study was supported by Maryland-CHART program. We would like to thank MDOT State Highway Administration and their staff, Egua Igbinosun, Mohammed Raqib, Jason Dicembre, and Sharon Hawkins, for their support.

Author Contributions

The authors confirm contribution to the paper as follows: study conception and design: Minsu Won, Hyeonmi Kim, Gang-Len Chang; data collection: Minsu Won, Hyeonmi Kim, Gang-Len Chang; analysis and interpretation of results: Minsu Won, Hyeonmi Kim, Gang-Len Chang; draft manuscript preparation: Minsu Won, Hyeonmi Kim, Gang-Len Chang. All authors reviewed the results and approved the final version of the manuscript.

References

1. Golob, T. F., W. W. Recker, and J. D. Leonard. An Analysis of the Severity and Incident Duration of Truck-involved Freeway Accidents. *Accident Analysis & Prevention*, Vol. 19, No. 5, 1987, pp. 375–395.
2. Giuliano, G. Incident Characteristics, Frequency, and Duration on a High Volume Urban Freeway. *Transportation Research Part A: General*, Vol. 23, No. 5, 1989, pp. 387–396.
3. Khattak, A. J., H. M. Al-Deek, and R. W. Hall. Concept of an Advanced Traveler Information System Testbed for the Bay Area: Research Issues. *Journal of Intelligent Transportation Systems*, Vol. 2, No. 1, 1994, pp. 45–71.
4. Khattak, A. J., J. L. Schofer, and M. H. Wang. A Simple Time Sequential Procedure for Predicting Freeway Incident Duration. *Journal of Intelligent Transportation Systems*, Vol. 2, No. 2, 1994, pp. 113–138.
5. Garib, A., A. E. Radwan, and H. Al-Deek. Estimating Magnitude and Duration of Incident Delays. *Journal of Transportation Engineering*, Vol. 123, No. 6, 1997, pp. 459–466.
6. El-Basyouny, K., and T. Sayed. Comparison of Two Negative Binomial Regression Techniques in Developing Accident Prediction Models. *Transportation Research Record: Journal of the Transportation Research Board*, 2006. 1950: 9–16.
7. Qi, Y., and H. Teng. An Information-based Time Sequential Approach to Online Incident Duration Prediction. *Journal of Intelligent Transportation Systems*, Vol. 12, No. 1, 2008, pp. 1–12.
8. Chung, Y. Development of an Accident Duration Prediction Model on the Korean Freeway Systems. *Accident Analysis & Prevention*, Vol. 42, No. 1, 2010, pp. 282–289.
9. Khattak, A., X. Wang, and H. Zhang. Incident Management Integration Tool: Dynamically Predicting Incident Durations, Secondary Incident Occurrence and Incident Delays. *IET Intelligent Transport Systems*, Vol. 6, No. 2, 2012, pp. 204–214.
10. Wang, X., S. Chen, and W. Zheng. Traffic Incident Duration Prediction Based on Partial Least Squares Regression. *Procedia-Social and Behavioral Sciences*, Vol. 96, 2013, pp. 425–432.
11. Junhua, W., C. Haozhe, and Q. Shi. Estimating Freeway Incident Duration Using Accelerated Failure Time Modeling. *Safety Science*, Vol. 54, 2013, pp. 43–50.
12. Hojati, A. T., L. Ferreira, S. Washington, and P. Charles. Hazard Based Models for Freeway Traffic Incident Duration. *Accident Analysis & Prevention*, Vol. 52, 2013, pp. 171–181.
13. Li, R., and P. Shang. Incident Duration Modeling Using Flexible Parametric Hazard-based Models. *Computational Intelligence and Neuroscience*, Vol. 2014, 2014, p. 33.
14. Zou, Y., K. Henrickson, D. Lord, Y. Wang, and K. Xu. Application of Finite Mixture Models for Analysing Freeway Incident Clearance Time. *Transportmetrica A: Transport Science*, Vol. 12, No. 2, 2016, pp. 99–115.
15. Khattak, A. J., J. Liu, B. Wali, X. Li, and M. Ng. Modeling Traffic Incident Duration Using Quantile Regression. *Transportation Research Record: Journal of the Transportation Research Board*, 2016. 2554: 139–148.
16. Li, X., A. J. Khattak, and B. Wali. Large-Scale Traffic Incident Duration Analysis: The Role of Multi-agency Response and On-Scene Times. *Transportation Research Record: Journal of the Transportation Research Board*, 2017. 2616: 39–48.
17. Ozbay, K., and P. Kachroo. *Incident Management in Intelligent Transportation Systems*. Artech House Inc., Mass., 1999.
18. Ozbay, K., and N. Noyan Estimation of Incident Clearance Times Using Bayesian Networks Approach. *Accident Analysis & Prevention*, Vol. 38, No. 3, 2006, pp. 542–555.
19. Boyles, S., D. Fajardo, and S. T. Waller. A Naive Bayesian Classifier for Incident Duration Prediction. Presented at 86th Annual Meeting of the Transportation Research Board, Washington, D.C., 2007.
20. Xiaoqiang, Z., L. Ruimin, and Y. Xinxin. Incident Duration Model on Urban Freeways Based on Classification and Regression Tree. *Proc., 2009 Second International Conference on Intelligent Computation Technology and Automation*, Changsha, Hunan, China, IEEE, New York, 2009.
21. Chang, H. L., and T. P. Chang. Prediction of Freeway Incident Duration based on Classification Tree Analysis. *Journal of the Eastern Asia Society for Transportation Studies*, Vol. 10, 2013, pp. 1964–1977.
22. Ma, X., C. Ding, S. Luan, Y. Wang, and Y. Wang. Prioritizing Influential Factors for Freeway Incident Clearance Time Prediction Using the Gradient Boosting Decision Trees Method. *IEEE Transactions on Intelligent Transportation Systems*, Vol. 18, No. 9, 2017, pp. 2303–2310.
23. Wang, W., H. Chen, and M. C. Bell. Vehicle Vreakdown Duration Modelling. *Journal of Transportation and Statistics*, Vol. 8, No. 1, 2005, pp. 75–84.
24. Wei, C. H., and Y. Lee. Sequential Forecast of Incident Duration Using Artificial Neural Network models. *Accident Analysis & Prevention*, Vol. 39, No. 5, 2007, pp. 944–954.
25. Valenti, G., M. Lelli, and D. Cucina. A Comparative Study of Models for the Incident Duration Prediction. *European Transport Research Review*, Vol. 2, No. 2, 2010, pp. 103–111.
26. Guan, L., W. Liu, X. Yin, and L. Zhang. Traffic Incident Duration Prediction Based on Artificial Neural Network. *Proc., 2010 International Conference on Intelligent Computation Technology and Automation*, Changsha, China, IEEE, New York, 2010.

27. Wu, W. W., S. Y. Chen, and C. J. Zheng. Traffic Incident Duration Prediction Based on Support Vector Regression. Proc., 11th International Conference of Chinese Transportation Professionals (ICCTP 2011): Towards Sustainable Transportation Systems, Nanjing, China, American Society of Civil Engineers (ASCE), 2011, pp. 2412–2421.
28. Vlahogianni, E. I., and M. G. Karlaftis. Fuzzy Entropy Neural Network Freeway Incident Duration Modeling with Single and Competing Uncertainties. *Computer Aided Civil and Infrastructure Engineering*, Vol. 28, No. 6, 2013, pp. 420–433.
29. Park, H., A. Haghani, and X. Zhang. Interpretation of Bayesian Neural Networks for Predicting the Duration of Detected Incidents. *Journal of Intelligent Transportation Systems*, Vol. 20, No. 4, 2016, pp. 385–400.
30. Park, H. J., S. Gao, and A. Haghani. Sequential Interpretation and Prediction of Secondary Incident Probability in Real Time. Presented at 96th Annual Meeting of the Transportation Research Board, Washington, D.C., 2017.
31. Lin, P.-W., N. Zou, and G.-L. Chang. Integration of a Discrete Choice Model and a Rule-based System for Estimation of Incident Duration: a Case Study in Maryland. Presented at 83rd Annual Meeting of the Transportation Research Board, Washington, D.C., 2004.
32. Kim, W., and G. L. Chang. Development of a Hybrid Prediction Model for Freeway Incident Duration: A Case Study in Maryland. *International Journal of Intelligent Transportation Systems Research*, Vol. 10, No. 1, 2012, pp. 22–33.
33. Ji, Y. Prediction of Freeway Incident Duration Based on the Multi-model Fusion Algorithm. *Proc., 2011 International Conference on Remote Sensing, Environment and Transportation Engineering (RSETE)*, Nanjing, China, IEEE, New York, 2011.
34. He, Q., Y. Kamarianakis, K. Jintanakul, and L. Wynter. Incident Duration Prediction with Hybrid Tree-based Quantile Regression. In *Advances in Dynamic Network Modeling in Complex Transportation Systems*, Springer, 2013, pp. 287–305.
35. Li, R., F. C. Pereira, and M. E. Ben-Akiva. Competing Risk Mixture Model and Text Analysis for Sequential Incident Duration Prediction. *Transportation Research Part C: Emerging Technologies*, Vol. 54, 2015, pp. 74–85.
36. Zhu, W., Z. Zeng, Y. Wang, and Z. Pu. Predicting Incident Duration based on Spatiotemporal Heterogeneous Pattern Recognition. Presented at 96th Annual Meeting of the Transportation Research Board, Washington, D.C., 2017.
37. Pettet, G., S. Nannapaneni, B. Stadnick, A. Dubey, and G. Biswas. Incident Analysis and Prediction Using Clustering and Bayesian Network. Proc., The First IEEE International Conference on Smart City Innovations (IEEE SCI 2017), San Francisco Bay Area, USA, Institute of Electrical and Electronics Engineers (IEEE), 2017.
38. Kim, W. Development of a Traffic Incident Management System for Contending with Non-recurrent Highway Congestion. Proc., 2014 IEEE Conference on Computer Vision and Pattern Recognition (CVPR), New York, NY, USA, Institute of Electrical and Electronics Engineers (IEEE) Computer Society, 2014.
39. Chang, G. L., and S. Rochon. *Performance evaluation and benefit analysis for CHART*. Coordinated Highways Action Response Team, Maryland Department of Transportation, 2016.
40. Agrawal, R., T. Imieli?ski, and A. Swami. Mining Association Rules between Sets of Items in Large Databases. Proc., 1993 ACM SIGMOD International Conference on Management of Data, Washington, D.C., Association for Computing Machinery (ACM), 1993, pp. 207–216.
41. Hahsler, M., B. Grün, and K. Hornik. A Computational Environment for Mining Association Rules and Frequent Item Sets. *Journal of Statistical Software*, Vol. 14, No. 15, 2005, pp. 1–25.

The Standing Committee on Freeway Operations (AHB20) peer-reviewed this paper (18-03263).

Article

TRR
JOURNAL OF THE TRANSPORTATION RESEARCH BOARD

Transportation Research Record
2018, Vol. 2672(14) 73–84

DOI: 10.1177/0361198118778941
journals.sagepub.com/home/trr
SAGE

Investigating and Modeling the Illegal U-Turn Violations at Medians of Limited Access Facilities

Omar Al-Sahili[1], Haitham Al-Deek[1], Adrian Sandt[1], Alexander V. Mantzaris[2], John H. Rogers, Jr[1], and Md Omar Faruk[1]

Abstract

Illegal U-turns on freeways and toll roads are risky maneuvers that sometimes result in the turning vehicles causing various types of collisions or disturbances to approaching traffic. These illegal U-turn maneuvers can occur at traversable grass medians and emergency crossovers. Limited literature was found regarding the impact of illegal U-turns on these facilities. Therefore, to understand the roadway and median characteristics that could influence drivers' propensity to commit illegal U-turns, a sequential modeling methodology was adopted. This methodology combined a Poisson regression model with a Least Absolute Shrinkage and Selection Operator (LASSO) regression procedure to predict the cited violations at traversable median segments. Additionally, a logistic regression model was developed to predict the probability of a cited violation at official use only emergency crossovers. These models included illegal U-turn citations and crashes for the Orlando and Miami metropolitan areas in Florida from 2011 to 2016. The findings indicated that the average distance between access points, median width, speed limit, segment length, and distance to nearest segment were significant in predicting cited violations at traversable medians. Furthermore, the distance to the nearest interchange, distance to the nearest adjacent crossover, and median width were significant in predicting the probability of a cited violation occurring at an emergency crossover. This study helps agencies to predict the locations of illegal U-turn violations and to prioritize roadways for possible treatment to minimize the potential risk of head-on or other collisions due to illegal U-turn events.

The risk of drivers attempting to make illegal U-turns on limited access facilities (freeways and toll roads) has not been studied or investigated thoroughly. Illegal U-turn movements occur when a driver turns across a dividing space between opposing lanes of travel (median section) or through a median opening (crossover) that is designated for official use only (*1*). These U-turns can sometimes result in vehicles losing control and crossing the median to the opposite direction of traffic. Analysis of citation data in Florida indicates that such illegal U-turns occur very frequently. Additionally, illegal U-turns can result in various types of crash, including rear-end, sideswipe, or severe head-on crashes, and cause disturbances to through traffic, significantly reducing capacity (*2*). Crashes can result from drivers losing control after performing the U-turn maneuver at a high speed or misjudging the gap and speed of opposing traffic, giving drivers limited time to react and avoid a crash. Limited research has analyzed the contributing factors that cause drivers to commit illegal U-turns on limited access facilities. Understanding the roadway characteristics that influence illegal U-turn violations can help officials predict these violations and allocate appropriate countermeasures to combat them. This can reduce crashes resulting from these illegal maneuvers and improve safety for all drivers.

The main goals of this research were to identify the contributing factors that might cause drivers to commit illegal U-turn violations on limited access facilities and predict the locations of these violations. Violation prediction models were developed that can determine the locations with the highest illegal U-turn violation risk and where to allocate countermeasures to combat these violations. These types of models have never been developed before.

This paper does not ignore the necessity of emergency crossovers on limited access facilities. Previous studies have considered the indirect damage resulting from limiting the access of emergency vehicles. Srinivasan et al. (*3*) stated that median barriers increase the response time of emergency vehicles. Considering that each minute of delay reduces the chances of survival from cardiac arrest by 7%–10% (*4*), the cost of delay of an emergency vehicle due to reducing life survival chances could be about \$640,000–\$910,000/minute

[1]Department of Civil, Environmental, and Construction Engineering, University of Central Florida (UCF), Orlando, FL
[2]Department of Statistics, University of Central Florida (UCF), Orlando, FL

Corresponding Author:
Address correspondence to Haitham Al-Deek: Haitham.Al-Deek@ucf.edu

Traversable Grass Median Section on State Road (SR) 408

Emergency Crossover on SR91

Figure 1. Images of a traversable grass median segment and an emergency crossover for official use only on toll roads in Florida (*8*).

[estimated life value is \$9.1 million (*5*)]. Therefore, this paper aims suggest ways to minimize the risks resulting from the misuse of emergency crossovers that are designated for official use only, while keeping in mind the necessity for these facilities.

This paper analyzed illegal U-turn events (citations and crashes) on limited access facilities in the Orlando (Central Florida) and Miami (South Florida) metropolitan areas from the years 2011 to 2016 (*6*). These two areas have the largest number of illegal U-turn activities in the state. Since it was not possible to collect data for every illegal U-turn violation at these locations (whether they were cited or not), only cited violations were used to identify factors leading to illegal U-turn activities and hotspots that may need treatment.

Illegal U-turn incidents were analyzed for two different types of median facility as shown in Figure 1: grass traversable median segments where drivers could cross over to the opposite direction of traffic and emergency crossovers for official use only, located on non-traversable median segments, such as physically separated medians (*7*).

Each type of illegal U-turn maneuver has unique characteristics and potential variables to be evaluated and modeled. To determine the significant factors that affect the number of illegal U-turn violations per mile and per year at the traversable grass median segments, a sequential modeling procedure that included a Poisson regression model followed by a Least Absolute Shrinkage and Selection Operator (LASSO) analysis method was used. For the emergency crossovers, a logistic regression model was developed to predict the probability of a citation at each crossover due to the limited number of emergency crossovers that had multiple citations during the study period.

Numerous potential explanatory roadway characteristics were analyzed and included in the developed models. The significant models developed here could be used as planning and evaluation tools for agencies to evaluate their median facilities both before and after construction of their roadway networks. To illustrate the potential of these models, an example application is presented. These models can help agencies determine the best locations to place emergency median crossovers or whether certain existing crossovers should be relocated.

Literature Review

Very limited studies have considered the risk of illegal U-turns on limited access facilities. However, several studies considered the operational and safety effects of U-turn maneuvers through median openings on partially controlled divided highways (*2*, *9–11*). Moreover, a few studies considered illegal U-turns as a cause of wrong-way driving (WWD) on limited access facilities. For example, Finley et al. listed illegal U-turns as a possible entry source for WWD (*12*), and proposed the use of a "Push Button Device" to control access points in low-volume areas (*13*), which could be potentially modified for freeway crossovers. In addition, Doctor mentioned that some WWD results from improper use of emergency openings (*14*), and the National Traffic Safety Board (NTSB) mentioned that illegal U-turns are considered a type of WWD (*15*). The design requirements and guidelines for emergency crossovers have been well documented in the iterature. The American Association of State Highway and Transportation Officials (AASHTO) design manual discussed several spacing requirements for emergency crossovers on rural freeways (*16*). Several states took the initiative and developed their own guidelines for the installation of median barriers and emergency crossovers on their freeways (*17*). For example, the Florida Department of Transportation (FDOT) Plans Preparation Manual guidelines includes additional spacing and location requirements (*18*), and the Texas Road Design Manual discusses design recommendations for crossovers that are as discouraging and inconspicuous as possible to the traveling public (*19*). The California Department of Transportation Highway Design Manual includes a section on

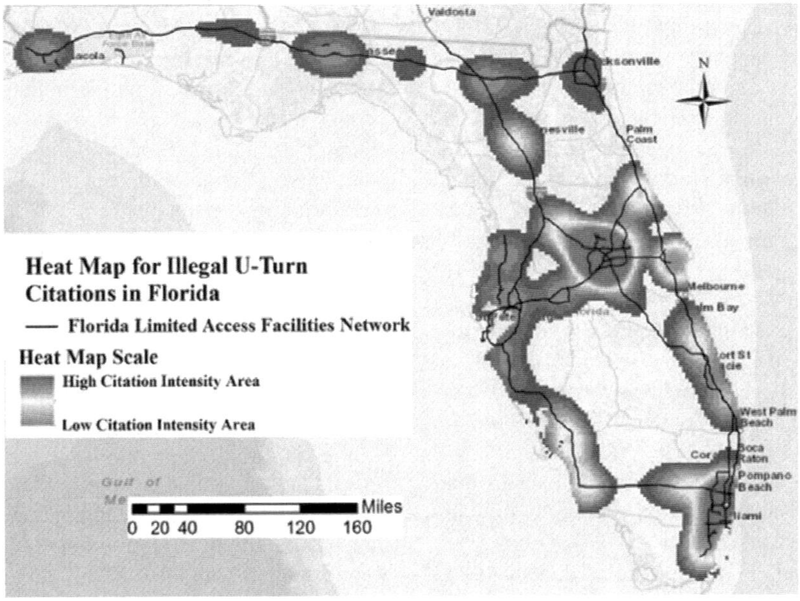

Figure 2. Heat map showing the illegal U-turn citations on limited access facilities network in Florida, 2011–2016 (map created by University of Central Florida [UCF]).

emergency crossovers as well (*20*). Furthermore, numerous studies investigated median crossover crashes and the success of median barriers in reducing such crashes (*21–25*), but minimal efforts were found on predicting and analyzing the contributing factors from the crossover crashes resulting from intentional illegal U-turn violations.

This literature review shows that there are existing design guidelines adopted by several states for median openings and crossovers. However, there is limited research on contributing factors that can be used to predict the locations of illegal U-turn maneuvers leading to several types of crashes on the roadway. This paper, therefore, investigates and analyzes the impact of roadway characteristics on illegal U-turn incidents and determines which of these characteristics create a high potential for violations. Previous studies have linked high WWD activity (such as WWD 911 calls and citations) on limited access facilities to WWD crash risk (*26–31*), so it is expected that reducing illegal U-turn violations could also reduce the crashes (or crash risk) resulting from such illegal maneuvers.

Methodology

The methodology in this paper has two major parts: data preparation and modeling. These are described in detail below.

Data Preparation

Citation Data. Illegal U-turn citation data from January 2011 to December 2016 for limited access facilities in Florida were analyzed to determine the high-risk locations of illegal U-turn violations (*26–31*). There was no unique Florida statute regarding illegal U-turn violations on limited access facilities. Therefore, two median crossover-related citation statutes were examined: 316.1515 (limitations on turning around) and 316.090(2) (median crossover on divided highways). A total of 620 illegal U-turn citations (240 at traversable median segments and 380 at emergency crossovers) were found after reviewing the data.

For the traversable median segments, most citations occurred during the afternoon period (noon–5:00 p.m.), and the months with the most citations were January, February, and May. For emergency crossovers, most citations occurred during the morning period (7:00–11:00 a.m.), and the months with the most citations were May, July, and September. Analysis of gender found that the percentage of female drivers who committed illegal U-turn maneuvers at paved emergency crossovers was significantly higher (at the 95% confidence level) than at traversable grass median segments.

Network Data. Median design and location of crossovers were examined on fully-controlled limited access facilities in the urban areas with high populations in Central and South Florida, as these two areas had the highest illegal U-turn citations statewide (Figure 2).

The median design was determined using aerial photos found in Google Earth (*8*), and for the Florida Turnpike Enterprise (FTE) network the aerial photos were verified with the as-built design plans provided by FTE. The considered traversable medians only included the median sections where the median did not have steep drainage or obstacles preventing drivers from accessing the opposite direction of traffic. In addition, 168 emergency crossovers were found in the study network: 73 in Central Florida and 95 in South Florida. The traversable medians and emergency crossovers installed in the study areas are shown in Figure 3.

The study network shown in Figure 3 includes the limited access facilities network in the urban areas with the highest population in Florida. However, the design recommendations proposed by AASHTO (*16*) and FDOT (*18*) are for median crossovers in rural areas. No specific guidelines were found for median crossovers in urban areas. Therefore, the rural area guidelines for median crossovers were considered for this research. However, additional guidelines for installing median barriers are adopted by the DOTs, but were not discussed in this paper.

In addition to the locations of traversable medians and crossovers, data on the number of police troopers assigned on each roadway were collected to have an appropriate exposure measure for the time and location of the citations. It was found that troopers' assignment was homogeneous on the studied roads.

Crash Data. Limited crashes resulting from illegal U-turn maneuvers were found in the areas studied. Therefore, it was decided to only perform the statistical modeling on the illegal U-turn citations. However, from 2011 to 2016, there were six crashes in the areas studied that resulted from illegal U-turn maneuvers (*32*). Reviewing these crash reports showed that two were head-on collisions, two were sideswipe collisions, and two were rear-end collisions. In addition, all crashes occurred in daylight and none of the drivers involved in these crashes was convicted with driving under influence. This suggests that drivers might be committing illegal U-turn maneuvers intentionally, possibly to take shortcuts, avoid paying tolls, or avoid excessive delays resulting from traffic congestion.

Modeling Methodology

Since crashes due to illegal U-turn maneuvers were rare, models were developed to predict cited violations in Central and South Florida. Two modeling approaches were adopted: one for cited violations on traversable median segments and one for cited violations at emergency crossovers. The first was a sequential analysis method that combined a Poisson regression model and the LASSO selection method to predict the number of cited violations on traversable median segments per mile per year, and the second was a logistic regression model to predict the probability of a cited violation occurrence each year at emergency crossovers located on non-traversable medians. The variables considered only included roadway design and traffic characteristics that the agencies could measure and control on their network, and possibly recondition after determining the locations with the highest risk due to illegal U-turn violations.

The dataset prepared for the traversable median model included 45 segments, where the number of cited U-turn violations located on each segment was measured as the response variable. Ten roadway design and traffic characteristics were considered as potential explanatory variables; these variables are shown and described in Table 1. A traversable segment is a continuous grass median that is not interrupted with any type of barrier or access restrictions and has similar traffic and roadway design characteristics. For the emergency crossover dataset, a total of 148 crossovers were included. The response variable was binary to represent the occurrence of a cited violation at the crossover. Eight roadway design and traffic variables were measured for each opening (Table 1).

All roadway design characteristics were measured manually using the aerial photos and measurement tools provided by Google Earth (*8*) and ArcGIS (*33*), and the traffic features were extracted from the FDOT Traffic Data website (*6*). The variables included in Table 1 were analyzed to find their influence on drivers' propensity to commit illegal U-turn violations.

Poisson and LASSO Regressions for Traversable Median Model

The developed cited violation prediction model for traversable segments predicts locations of illegal U-turn violations cited on these segments and determines a set of significant variables ranked by their effect and significance in predicting the cited violations. This helps guide agencies to where and how they can reduce these violations. To develop this model, a sequential analysis technique was performed by conducting a Poisson regression that models the expected number of citations per year and determines the significant explanatory variables, followed by a LASSO selection methodology for the subset of significant variables found by Poisson. The Poisson regression model was selected to support the LASSO method due to its proven efficiency in modeling rare events over a certain time period, like the citations analyzed in this study (*34*). LASSO has not been used in analyzing traffic citations before. However, it was selected in this research for its abilities to exclude nuisance variables with little predictive contributions and its efficiency in evaluating several independent variables and sorting them by their significance to determine the variables with the highest priority for agencies to consider (*35*). Furthermore, LASSO could model rate outcome variables instead of only modeling discrete outcome variables as in the Poisson model. The outcome produced from the LASSO model was the number of cited illegal U-turn violations per mile per year.

The correlation between all potential explanatory variables was tested, and a moderate correlation was found

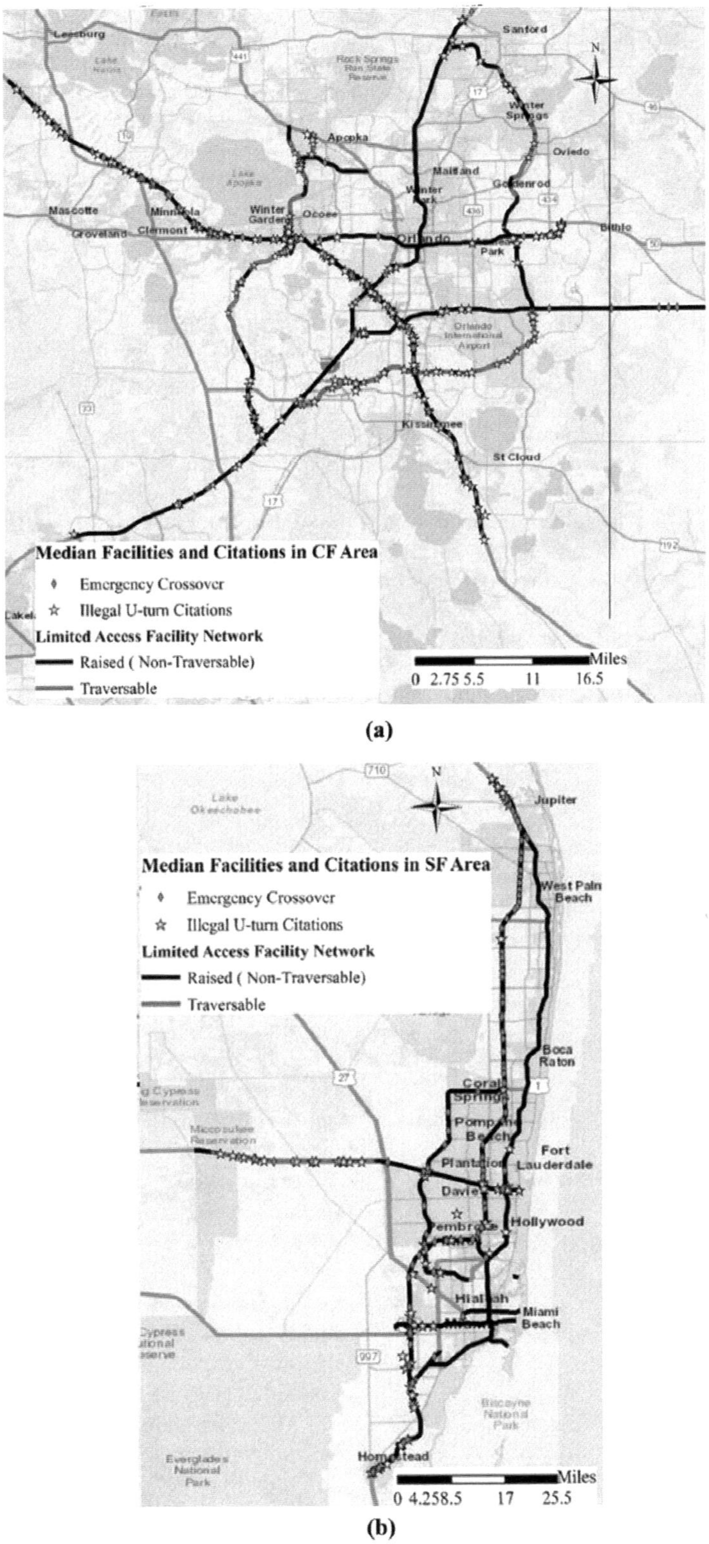

Figure 3. Study areas with citations, locations of median openings (crossovers), and traversable median segments (maps created by UCF): (*a*) Central Florida (CF); (*b*) South Florida (SF).

Table 1. Variable Descriptions for the Traversable Medians and Emergency Crossovers Models

Variable	Description of variable	Mean	Variance
No. of citations	No. of citations/year in each segment (for Poisson model)	3.49	7.46
Rate of citations	No. of citations/mile/year in each segment (for LASSO model)	0.57	1.67
Citations occurrence	If a citation occurred at the crossover (Yes = 1, No = 0)	0.32	0.47
Segment length	Length of traversable median segment [mile]	1.91	5.31
Distance to nearest interchange	Distance from center of segment to nearest major interchange [mile]	2.41	8.29
No. of lanes	Number of lanes in each direction	2.29	0.30
No. of access points	Number of access points in the segment	1.62	1.51
Crossover presence	If an emergency crossover is in the segment [1 = Yes, 0 = No]	0.38	0.24
Average distance per access point	The average distance between access points in the segment [mile]	1.59	1.97
Distance to the nearest traversable segment	The distance to the closest traversable median segment [mile]	0.86	2.08
Median width	Median width [Ft]	79.65	47.37
Average annual daily traffic (AADT)	Average annual daily traffic on segment over study period [×1000]	75.89	73.77
Speed limit	Posted speed limit	68.44	7.75
Distance to interchange	Distance from the opening to the nearest major interchange	4.36	5.52
Distance to the nearest crossover	Distance from a crossover to the next adjacent crossover [mile]	1.39	1.07
Distance to toll plaza	Distance to the nearest mainline toll plaza [mile]	10.67	16.69
Distance to access point	Distance to the nearest on-ramp or off-ramp [mile]	2.62	3.73
No. of lanes	Number of lanes	2.70	0.77
Median width	Median width [Ft]	42.72	31.67
AADT	Average annual daily traffic on segment over study period [×1000]	75.76	1558.45
Speed limit	Posted speed limit	70	7.78

between the AADT and the number of lanes in the model (correlation coefficient of 0.578) (*36*). However, neither of these two variables was found significant in the final model.

To determine the starting significant variables, all variables were included in the Poisson model following the general form:

$$Y_i = E(Y_i) + \varepsilon, i = 1, 2, ..., n \quad (1)$$

where Y_i is the non-negative response variable with mean μ that depends on $k-1$ predictor variables X_1 through X_{k-1} as shown in the following equation, and ε is the error value:

$$\mu_i = \exp(\beta_0 + \sum_{j=1}^{k} \beta_j X_j) \quad (2)$$

where μ_i is the mean of Y_i for the ith set of values for predictor variables $X_1, ..., X_{k-1}$

β_0 = Intercept
k = Number of significant predictor variables
X_j = Explanatory variable j
β_j = Estimated coefficient for x_j
exp = Exponential function

The Poisson regression assumes that the counts follow the Poisson distribution. To account for potential differences between the actual distribution and the Poisson distribution, a scale parameter was introduced using SAS software to measure the distribution and fit the overdispersion in the response variable (*37*). The scale value was 1.0798, meaning the standard errors in the model are multiplied by $\sqrt{1.0798}$ to fit the Poisson distribution. The closer the scale is to 1 the less the data is overdispersed.

The results from the Poisson regression modeling found that seven of the 10 analyzed variables were significant with a probability <.05 that the Wald x^2 statistic for each variable was larger than the critical x^2 for 38 degrees of freedom. The non-significant variables were the number of lanes, the presence of emergency crossover, and the number of access points in the traversable segment. The variables chosen by the Poisson model and their statistics are shown in the first five columns of Table 2.

Next, the LASSO variable selection method was used on the significant variables chosen by the Poisson model. LASSO minimizes the coefficients of the ineffective variables to minimize the sum of squared errors (SSE) estimate of the model and accordingly increases the prediction accuracy of the model (*35*). The SSE is the difference between the actual observations and the predicted values from the model. Furthermore, the LASSO method defines a shrinkage parameter (constraint) that determines the number of variables and the level of significance for the variables to be included in the model (*35*). The shrinkage parameter is defined as [|*b*|/Final |*b*|], where |*b*| is the SSE in the model. LASSO produces a generalized linear model that predicts the number of citations per mile per year, as

Table 2. Comparison between the Poisson and the LASSO Significant Models

Variables	Coefficients of Poisson model	Standard error	X^2	Pr > x^2	Order included by LASSO	Coefficients of LASSO model	Adjusted R^2 for LASSO model
Intercept	6.0574	2.3648	6.56	.0104	0	6.6984	0
Average distance between access point	0.4665	0.0387	133.77	<.0001	1	0.6293	0.0937
Median width	0.0052	0.0019	7.35	.0075	2	0.0182	0.5028
Speed	−0.0849	0.0351	5.46	.0195	3	−0.1202	0.5980
Segment length	0.1379	0.0471	8.09	.0034	4	−0.0619	0.6103
Distance to nearest segment	−0.4441	0.1269	17.81	<.0001	5	−0.1085	0.6267*
Distance to nearest interchange	−0.1529	0.0616	7.58	.0059	6	Not selected	0.6260
AADT	−0.0045	0.0027	0.07	.0677	7	Not selected	0.6165

*Adjusted R^2 marked with an asterisk is the maximum adjusted R^2 value reached in the model (at step 5).

$$\text{Rate}\ (Y) = \beta_0 + \beta_1 X_1 + \beta_2 X_2 + \ldots + \varepsilon \quad (3)$$

where:

Rate (Y): Number of illegal U-turn citations per mile per year at each segment
β_n: Coefficients of model variables, where n = 0 to n
X_n: Variables included in the model, where n = 1 to n
ε: Error term.

The results of this model are shown in the last three columns of Table 2. It should be noted that the two models (Poisson and LASSO) have different modeling formulas; therefore, the values of the coefficients are not comparable. However, the coefficients' signs are comparable and have similar effect on the response variable. Compared with the Poisson model, the LASSO procedure also shows which order the variables entered the model based on their significance. A coefficient development and average squared error illustration that shows the effect of the variables included in the model is shown in Figure 4. The standardized coefficient is shown in the y-axis of the upper illustration of Figure 4a, which demonstrates the effect of the variables included in the model. The shrinkage parameter is the x-axis in Figure 4, a and b. The impact of including each variable in the model on the adjusted R^2 value is shown in Figure 4a and the average squared error for the model is shown in Figure 4b, both are goodness of fit measures of the model.

As shown in Table 2, the distance to the nearest interchange and AADT variables were not selected in the LASSO model because these two variables would reduce the performance of the model. The average distance between access points was the most significant variable, indicating that the presence of on-ramps or off-ramps far from each other would encourage drivers to commit an illegal U-turn to reduce the required travel time if needing to correct their route. The next most significant variable was the median width, indicating that large medians provide storage and a feeling of protection for turning vehicles. The third variable was the speed limit; lower speed limits can make it easier for drivers to slow down to commit the illegal U-turn maneuver. In the LASSO model, it is shown that shorter continuous flushed segments encourage drivers to perform illegal U-turns and would have higher citations per mile than longer segments. In the Poisson model, which predicts citations per segment instead of citations per mile, longer segments have a larger opportunity for committing more violations. The distance to the nearest segment variable indicates that segments located closer to each other would be more likely to attract violations. In addition the SSE for the LASSO model was 47.4 compared with 147.9 for the Poisson model, indicating that LASSO has less error in its prediction.

Logistic Regression for Emergency Crossover Model

A logistic regression model was used to find the probability of an illegal U-turn citation occurring at an emergency crossover located on non-traversable median segments. The logistic regression produces a binary outcome that is either 1 (event) if a citation will probably occur or 0 (non-event) if a citation will probably not occur. The model equation is shown below:

$$P(x) = \frac{1}{1 + \exp{-(\beta_0 + \beta_1 X_1 + \beta_2 X_2 \ldots)}} \quad (4)$$

Where:

$P(x)$: The probability of a citation occurring at the median opening
exp: Exponential function
β_n: Coefficients of model variables, where n = 0 to n
X_n: Variables included in the model, where n = 1 to n

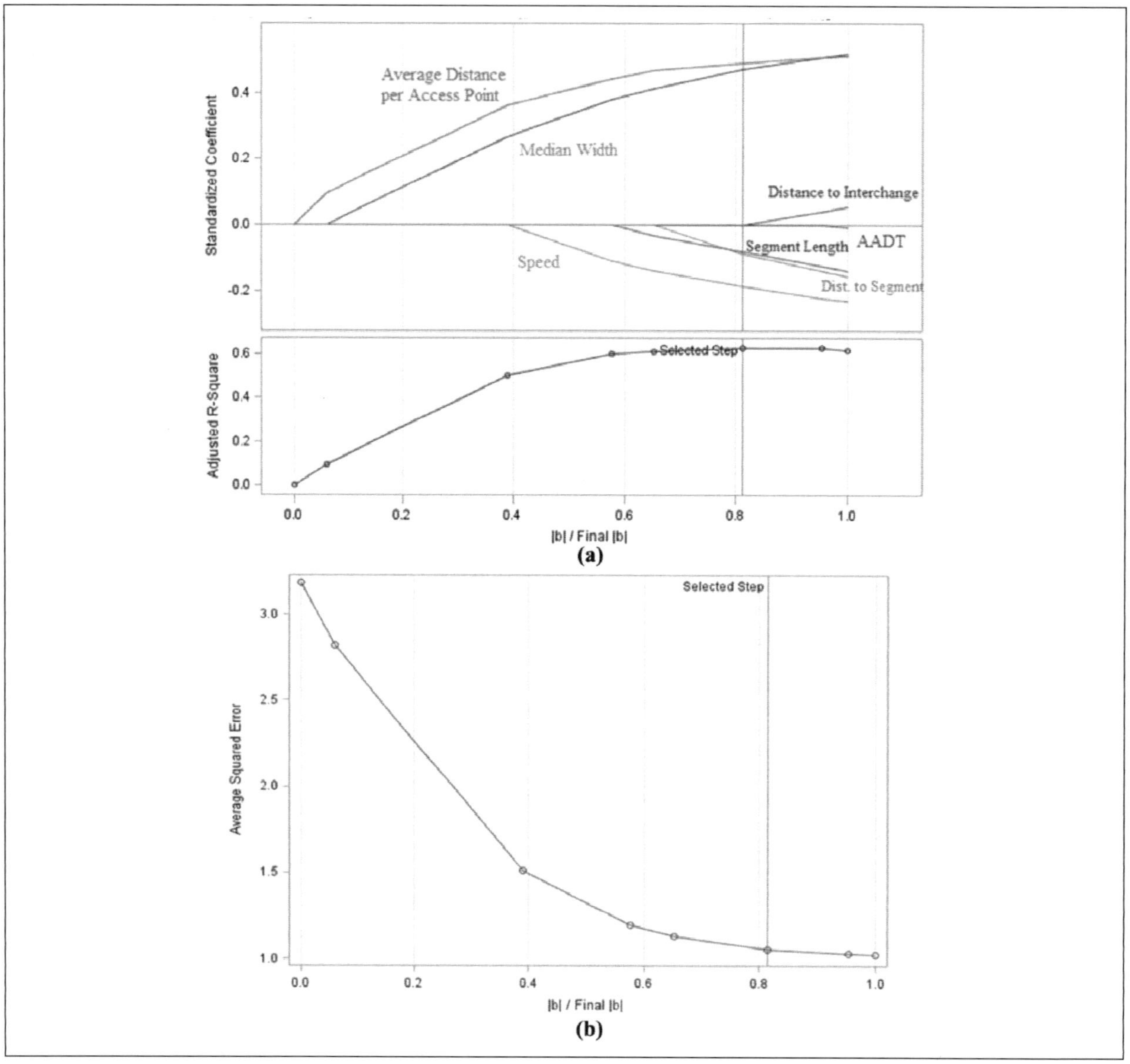

Figure 4. LASSO coefficient development and average squared errors illustrations for the variables included in the model: (*a*) LASSO coefficient development illustrations for the variables in the model; (*b*) average squared errors for the variables in the model.

The correlation was checked for the variables included. AADT and number of lanes variables had a high correlation coefficient of 0.735 (*36*). However, neither variable was found significant in the final model.

A stepwise selection procedure that included the variables with 90% significance level was performed. This resulted in three significant variables being included: distance to the nearest interchange, distance to the nearest crossover, and median width, see Table 3.

This model is accurate at predicting whether a citation will occur, as the proportion of observations estimated correctly by the model was $c = 0.727$. The model estimates indicate that emergency crossovers near interchanges, near other crossovers, or installed on wide medians have increased probability of violations occurring. The presence of multiple crossovers could make these crossovers more noticeable to drivers. Emergency crossovers near interchanges could be used by drivers who make a wrong exit or who want to avoid paying tolls at tollbooths located at ramps. Large median widths would provide storage and large turning radii which would encourage drivers to commit illegal U-turns maneuvers.

Example for Model Implementation

The models described in this paper can be used as evaluation tools in the planning and design of freeway facilities. They

Table 3. Model Estimates for Violations on Emergency Crossovers

Variable	Coefficient	Standard error	Upper 95% CI	Lower 95% CI	Wald X^2	Pr > X^2
Distance to nearest interchange [miles]	−0.1916	0.0862	−0.3606	−0.0226	4.9360	.0263**
Distance to nearest crossovers [miles]	−0.2517	0.1525	−0.5506	0.0472	2.7246	.0988*
Median width [Ft]	0.0123	0.00709	−0.0016	0.0262	2.9903	.0838*
Intercept	0.3117	0.5350	−0.7369	1.3604	0.3395	.5601

Note: CI = confidence interval.
*90% significance level; **95% significance level.

can be used to determine locations with low potential violation risk for installing new or relocating existing median openings, or to find locations with the highest probability of illegal U-turns for installing countermeasures. For example, an agency constructing a new highway or installing median barriers on their roadway network can use these tools to determine the locations where emergency crossovers can be installed. As a first step, agencies should follow AASHTO and state guidelines for spacing, location, and other installation requirements (*16*). Afterwards, they can use the model to find locations with the lowest violation probabilities.

The example shown in Figure 5 illustrates a South Florida section on I-75 near its interchange with SR-869. Figure 5 shows an existing crossover on this section with two suggested alternatives for possible relocation of this crossover.

The three roadway design characteristics shown in Table 3 were measured for the existing crossover and for each of the suggested two alternative locations (options A and B) shown in Figure 5. Equation 5 was then used to find the probability of illegal U-turn violations at each location.

$$P(x)=\frac{1}{1+\exp-\begin{pmatrix}0.3117-0.1916*Dist.to\ Interchange\\ -0.2517*Dist.\ too\ Cossover\\ +0.123*Median\ Width\end{pmatrix}} \quad (5)$$

The probability of a violation occurring at the existing crossover location is 77%, whereas the probabilities of violation occurring at options A and B are 39% and 47% respectively. Taking into consideration that both options follow the recommended spacing guidelines and requirements, option A is a better location for installing an emergency crossover on this section because it will reduce the probability of illegal U-turn violations by 38% compared with the existing location.

This example, which demonstrated only one possible application scenario, shows how the crossover model could be used in evaluating the existing conditions at already installed openings to determine the locations with the lowest and highest probability of violations and consequently the lowest and highest risk due to illegal U-turns. In addition, the traversable median models can be applied in other scenarios where locations with high or low expected illegal U-turn violations per mile can be determined to select hotspot locations for treatments or median barrier installations.

Summary and Conclusions

There has been a lack of research into crashes and citations resulting from illegal U-turn violations (including head-on due to median crossover, run-off the road, sideswipe, and rear-end crashes). If left unaddressed, these violations could increase with the increase of traffic congestion nationwide. To help reduce future traffic crashes with fatalities and or severe injuries that could result from illegal U-turn violations, this paper developed prediction models for high risk locations.

The paper analyzed the illegal U-turn violations cited at traversable median sections (flushed grass median) and at emergency crossovers in both Central and South Florida areas, the two areas with the highest illegal U-turn violations in Florida. A sequential modeling approach that included both Poisson regression and the LASSO variable selection method was adopted for the traversable median segment. This approach found that the average distance between access points, the median widths, the speed limits, the segment length, and the distance to the nearest segment were the most significant variables in predicting illegal U-turn violations at traversable median segments. For the emergency crossovers, a logistic regression model was used to predict the probability of a violation occurrence. This model found that crossovers near major interchanges or other crossovers and ones installed at wide median segments had the highest probability of illegal U-turn violations, which could increase the risk of crashes resulting from the illegal U-turn maneuvers.

This paper suggested a new evaluation approach for medians on limited access facilities that could be considered by agencies in their decision-making process for installing and/or evaluating median facilities and prioritizing their highways for treatment by placing proper countermeasures that could reduce or eliminate illegal U-turns. Moreover, the significant contributing factors found in this paper can be considered in determining the targeted roadway features for treatment. An example application was provided to show how agencies can analyze emergency crossovers using the developed model.

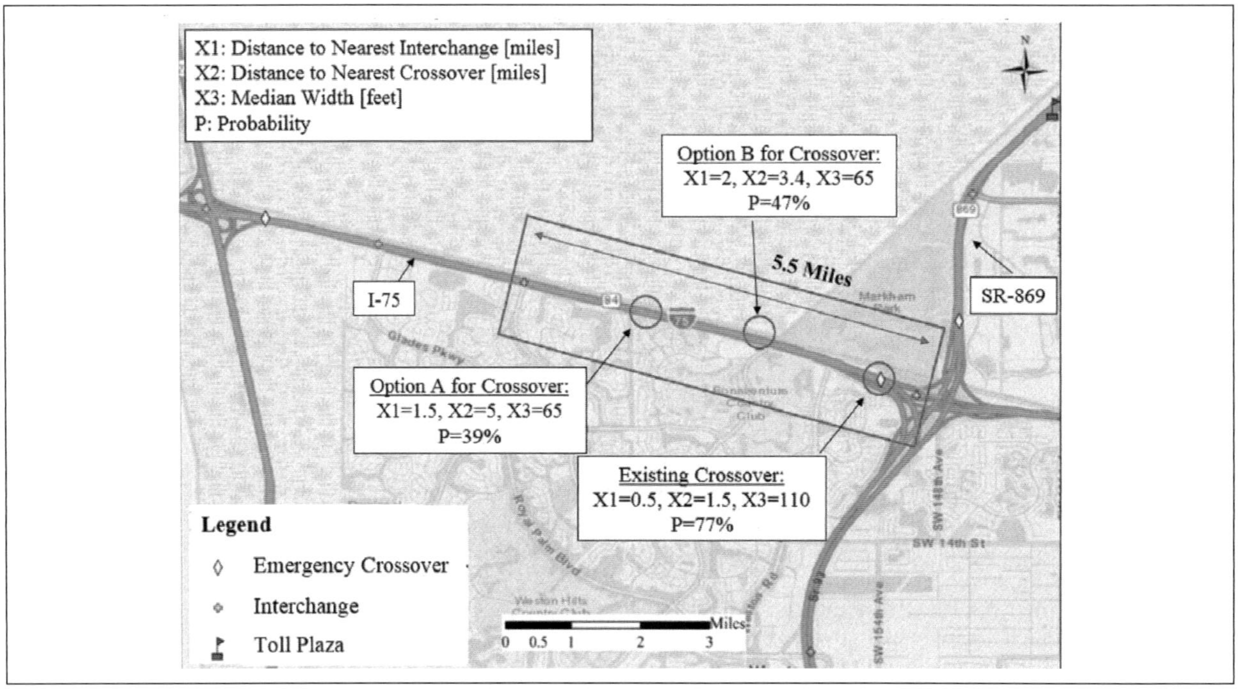

Figure 5. Application example of the emergency crossover location model.

For locations that have high illegal U-turn violations, treatment countermeasures can include additional law enforcement, additional warning signs, surveillance cameras, and automatically operated gates for official use only. The best countermeasures will vary depending on the characteristics of each location. Future research could examine these different countermeasures to determine their effectiveness in reducing illegal U-turn citations and any potential negative consequences to emergency response times. Moreover, further microscopic investigation could be conducted to find more precise factors that might encourage drivers to commit these illegal U-turn maneuvers. The research methodology developed in this paper could also be expanded and transferred to other states and regions. Using these approaches for other divided highways in Florida and other states can help reduce these illegal U-turn violations, which will improve safety and traffic operations.

Acknowledgments

The research discussed in this paper was funded by the Central Florida Expressway Authority and the Florida Turnpike Enterprise. In addition, the authors would like to acknowledge the help of Grady Carrick of Enforcement Engineering, Inc.

Author Contributions

The authors confirm contribution to the paper as follows: study conception and design: Al-Sahili, Al-Deek, Sandt, Mantzaris, Rogers, and Faruk; data collection: Al-Sahili, Al-Deek, Sandt, Mantzaris, Rogers, and Faruk; analysis and interpretation of results: Al-Sahili, Al-Deek, Sandt, Mantzaris, Rogers, and Faruk; draft manuscript preparation: Al-Sahili, Al-Deek, Sandt, Mantzaris, Rogers, and Faruk. All authors reviewed the results and approved the final version of the manuscript.

References

1. The Florida Legislature Website. *The 2016 Florida Statutes*. www.leg.state.fl.us/Statutes/index.cfm?App_mode=Display_Statute&Search_String=&URL=0300-0399/0316/Sections/0316.090.html. Accessed May 20, 2017.
2. Haleem, K., and M. Abdel-Aty. Association Between Access Management and Traffic Safety: Median Classification and Spatial Effect. *ITE Journal: Institute of Transportation Engineers*, Vol. 82, No. 4, 2012, pp. 22–27.
3. Srinivasan, R., K. Lacy, J. Feaganes, and W. Hunter. *Effects of Continuous Median Barriers on Highway Speeds, Emergency Response Times, and Transport Times on North Carolina Highways*. Report FHWA-NC-2003-05. University of North Carolina Highway Safety Research Center, Chapel Hill, NC, 2003.
4. FHWA. *The Traffic Signal Preemption for Emergency Vehicles: A cross-cutting study*. FHWA. https://ntl.bts.gov/lib/jpodocs/repts_te/14097_files/14097.pdf. Accessed June 13, 2017.
5. *Guidance on Treatment of the Value of Preventing Fatalities and Injuries in Preparing Economic Analyses: 2015 Adjustment*. Memorandum from U.S. Department of Transportation. June 17, 2015. https://cms.dot.gov/sites/dot.gov/files/docs/VSL2015_0.pdf. Accessed July 12, 2017.

6. Florida Department of Transportation. *Traffic Data Shapefiles: Transportation Data and Analytics Office*. http://www.fdot.gov/planning/statistics/gis/trafficdata.shtm. Accessed May 11, 2017.
7. FHWA. *The Manual on Uniform Traffic Control Devices (MUTCD)*. U.S. Department of Transportation, FHWA, Washington, D.C., 2012.
8. Google Earth. *Florida Map*. https://earth.google.com/web/@27.62378335,-81.5076548,-29.53734374a,1297823.57209772d,35y,-0h,0t,0r. Accessed July 10, 2017.
9. Levinson, H., D. Potts, J. Harwood, and D. Torbic. Safety of U-Turns at Unsignalized Median Openings: Some Research Findings. *Transportation Research Record: Journal of the Transportation Research*, 2005. 1912: 72–81.
10. Potts, I. B., D. W. Harwood, D. J. Torbic, K. Richard, J. Gluck, H. S. Levinson, P. Garvey, and R. Ghebrial. *NCHRP Report 524: Safety of U-turns at Unsignalized Median Openings*. Transportation Research Board of the National Academies, Washington, D.C., 2004.
11Xu, L., and B. Daniel. Fambro Student Paper Award: Right Turns Followed by U-Turns Vs. Direct Left Turns: A Comparison of Safety Issues. *ITE Journal: Institute of Transportation Engineers*, Vol. 71, No. 11, 2001, pp. 36–43.
12.Finley, M. D., S. P. Venglar, V. Iragavarapu, J. D. Miles, E. S. Park, S. A. Cooner, and S. E. Ranft. *Assessment of the Effectiveness of Wrong Way Driving Countermeasures and Mitigation Methods*. Report FHWA-TX-15/0-6769-1. Texas A&M Transportation Institute, College Station, TX, 2014.
13. Finley, M. D., P. Songchitruksa, and S. R. Sunkari. *Evaluation of Innovative Devices to Control Traffic Entering from Low-Volume Access Points Within a Lane Closure*. Report FHWA/TX-13/0-6708-1. Texas A&M Transportation Institute, College Station, Texas, 2013.
14. Doctor, M. Wrong Way Driving: New Focus on a Persistent Problem. Presentation for TRB Webinar, Transportation Research Board of the National Academies, Washington, D.C., 2016.
15.National Transportation Safety Board. *Wrong-Way Driving. Highway Special Investigation Report*. NTSB/SIR-12/01, 2012. https://www.ntsb.gov/safety/safety-studies/Documents/SIR1201.pdf. Accessed July 13, 2017.
16. AASHTO. *A Policy on Geometric Design of Highways and Streets*, 6th ed. AASHTO, Washington, D.C., 2011.
17.CTC & Associates LLC. *Putting the Brakes on Crossover Crashes: Median Barrier Research and Practice in the U.S.* March 30, 2007. http://research.transportation.org/Documents/tsrmedianbarriers.pdf. Accessed June 15, 2017.
18. *Plans Preparation Manual 2017*. Florida Department of Transportation. http://www.fdot.gov/roadway/ppmmanual/2017/Volume1/2017Volume1.pdf. Accessed February 17, 2017.
19. *Roadway Design Manual*. Texas Department of Transportation. http://onlinemanuals.txdot.gov/txdotmanuals/rdw/rdw.pdf. Accessed May 17, 2017.
20. *Caltrans Highway Design Manual*. California Department of Transportation. http://www.dot.ca.gov/hq/oppd/hdm/pdf/english/HDM_Complete_07Mar2014.pdf. Accessed July 15, 2017.
21.Sicking, D., F. de Albuquerque, K. Lechtenberg, and C. Stolle. Guidelines for Implementation of Cable Median Barrier. *Transportation Research Record: Journal of the Transportation Research Board*, 2009. 2120: 82–90.
22. Russo, B., P. Savolainen, and T. Gates. Development of Crash Modification Factors for Installation of High-Tension Cable Median Barriers. *Transportation Research Record: Journal of the Transportation Research Board*, 2016. 2588: 116–125.
23. Villwock, N., N. Blond, and A. Tarko. Cable Barriers and Traffic Safety on Rural Interstates. *Journal of Transportation Engineering: ASCE*, Vol. 137, No. 7, 2011, pp. 248–259. https://doi.org/10.1061/(ASCE)TE.1943-5436.0000213
24. Marzougui, D., C. Kan, and K. Opiela. Further Considerations for Effective Median Barrier Lateral Placement for Varying Highway Cross Sections. *Transportation Research Record: Journal of the Transportation Research Board*, 2014. 2437: 63–77.
25.Alluri, P., K. Haleem, A. Gan, and J. Mauthner. Safety Performance Evaluation of Cable Median Barriers on Freeways in Florida. *Traffic Injury Prevention*, Vol. 17, No. 5, 2016, pp. 544–551. https://doi.org/10.1080/15389588.2015.1101079
26. Sandt, A., H. Al-Deek, and J. H. Rogers. Identifying Wrong-Way Driving Hotspots by Modeling Crash Risk and Analyzing Traffic Management Center Response Times. *Transportation Research Record: Journal of the Transportation Research Board,* 2017. 2616: 58–68.
27. Rogers, J. H., A. Sandt, H. Al-Deek, A. H. Alomari, N. Uddin, E. Gordin, C. D. Santos, J. Renfrow, and G. Carrick. Wrong-Way Driving: Multifactor Risk-Based Model for Florida Interstates and Toll Facilities. *Transportation Research Record: Journal of the Transportation Research Board*, 2015. 2484: 119–128.
28.Rogers, J. H., H. Al-Deek, A. H. Alomari, E. Gordin, and G. Carrick. Modeling the Risk of Wrong-Way Driving on Freeways and Toll Roads. *Transportation Research Record: Journal of the Transportation Research Board*, 2016. 2554: 166–176.
29.Rogers, J. H., H. Al-Deek, A. Alomari, F. A. Consoli, and A. Sandt. Wrong-Way Driving on Florida Toll Roads: An Investigation into Multiple Incident Parameters and Targeted Countermeasures for Reductions. *International Journal of Engineering Management and Economics*, Vol. 5, No. 3–4, 2015, pp. 144–168.
30.Rogers, J. H. *Evaluating Wrong-Way Driving for Florida Interstates and Toll Road Facilities: A Risk-Based Investigation, and Countermeasure Development*. PhD dissertation. University of Central Florida, Orlando, Florida, 2016.
31. Rogers, J. H., A. Sandt, H. Al-Deek, and A. Alomari. Wrong-Way Driving Multifactor Prediction Models for Florida Limited Access Facilities. Presented at Road Safety and Simulation International Conference 2015 RSS2015, Orlando, Florida, 2015.
32. The GeoPlan Center at the University of Florida. *Signal Four Analytics Website*. https://s4.geoplan.ufl.edu. Accessed June, 2017.
33. ESRI Inc. *ArcGIS. ArcGIS 10.3.1 for Desktop*, 2015.
34.Li, Z., W. Wang, P. Liu, J. M. Bigham, and D. Ragland. Using Geographically Weighted Poisson Regression for County-Level Crash Modeling in California. *Safety Science*, Vol. 58, 2013, pp. 89–97. https://doi.org/10.1016/j.ssci.2013.04.005
35. Tibshirani, R. Regression Shrinkage and Selection via the Lasso. *Journal of the Royal Statistical Society. Series B:*

Methodological, Vol. 58, 1996, pp. 267–288. http://www.jstor.org/stable/2346178
36. Evans, J. D. *Straightforward Statistics for the Behavioral Sciences*. Thomson Brooks/Cole Publishing Co., Belmont, CA, 1996.
37. SAS Institute Inc. *SAS 9.4*. SAS Institute Inc., Cary, NC, 2012.

The Standing Committee on Freeway Operations (AHB20) peer-reviewed this paper (18-03095).

This paper and its contents, including conclusions and results, are solely those of the authors; they do not represent opinions or policies of the funding agencies.

TRR
JOURNAL OF THE TRANSPORTATION RESEARCH BOARD

Article

Transportation Research Record
2018, Vol. 2672(14) 85–95

DOI: 10.1177/0361198118778938
journals.sagepub.com/home/trr
SAGE

A Wrong-Way Driving Crash Risk Reduction Approach for Cost-Effective Installation of Advanced Technology Wrong-Way Driving Countermeasures

Adrian Sandt[1] and Haitham Al-Deek[1]

Abstract

Wrong-way driving (WWD) is hazardous on high-speed limited access facilities. Traditional signage and pavement markings will not always prevent intoxicated or confused drivers from entering these facilities the wrong way. To better alert wrong-way drivers, agencies can consider WWD countermeasures equipped with advanced technologies (including warning lights and detection devices) on exit ramps. However, these countermeasures can be expensive for agencies to install on entire roadways or corridors. This paper develops an innovative WWD crash risk (WWCR) reduction approach consisting of a WWCR segment model and an optimization algorithm that can be used to help agencies decide where to install WWD countermeasures. The approach examines segments of limited access facilities to determine the interchanges where advanced technology WWD countermeasures will provide the greatest reduction in WWCR based on an agency's available resources. A hypothetical example application of this approach is shown for the Central Florida Expressway Authority toll road network. This example shows how the WWCR reduction approach can help agencies identify the optimal investment level and high-risk locations. It also shows how the optimization algorithm can provide significant cost savings compared with equipping entire roadway segments (57% savings) or corridors (83% savings). Agencies can customize the algorithm by adding constraints to represent various scenarios and make the algorithm applicable for networks ranging from single roadways to statewide systems. This WWCR reduction approach could be utilized by agencies nationwide to help them save resources and prioritize investment.

Wrong-way driving (WWD) can be difficult for transportation agencies to combat, as it is uncommon and often uncertain where the wrong-way driver entered the roadway. However, WWD crashes typically have severe consequences, especially on high-speed limited access facilities. Traditional WWD countermeasures, such as "Wrong Way" signs and pavement arrows, are commonplace through the United States, but these are sometimes not enough. Intoxicated, confused, or elderly wrong-way drivers might not notice these countermeasures, especially at night. To alert these unintentional wrong-way drivers, countermeasures with advanced technologies might be needed. These countermeasures can include signs with flashing lights, radar detectors, and cameras. One example of such a countermeasure is "Wrong Way" signs equipped with Rectangular Flashing Beacons (RFBs) that flash when a wrong-way driver is detected. Compared with traditional WWD countermeasures, advanced technology WWD countermeasures could cause more wrong-way drivers to turn around before they enter an access-controlled roadway, but they also cost more.

Due to limited resources, agencies probably cannot afford to install advanced technology WWD countermeasures at all exit ramps on their limited access facilities at once. To help agencies decide where these countermeasures should be installed first based on available resources, this research has developed a new WWD crash risk (WWCR) reduction approach. This approach consists of a model to determine WWCR on roadway segments and an optimization algorithm to help agencies decide on the best interchanges to equip with countermeasures for a set investment amount. With this approach, agencies can identify which interchanges will provide the most cost-effective WWCR reduction (maximum WWCR reduction for the lowest cost). By itself, the WWCR reduction approach is not a decision-making

[1]Department of Civil, Environmental, and Construction Engineering, University of Central Florida (UCF), Orlando, FL

Corresponding Author:
Address correspondence to Haitham Al-Deek: Haitham.Al-Deek@ucf.edu

tool, but a way to provide more information to agencies before they decide on where to install countermeasures.

To illustrate the capabilities and application of this WWCR reduction approach, a hypothetical example is provided. This example uses results from the WWCR segment modeling in Central Florida discussed in (*1*), which predicted the number of WWD crashes in five-interchange segments of limited access facilities over a four-year period based on WWD citations, WWD 911 calls, interchange design, and traffic volumes. The actual and predicted WWD crashes were added together to obtain a WWCR value for each segment. This WWCR value considers actual WWD crashes and the potential for WWD crashes due to non-crash WWD events and other significant factors. Ranking segments by WWCR can help agencies identify segments with high potential for WWD crashes (WWD hotspots) rather than just relying on WWD crashes alone (*1*).

While the WWCR model identifies WWD hotspot segments, it is not able to identify individual interchanges. Therefore, agencies would have to equip all exit ramps in the hotspot segments to achieve the most WWCR reduction (*1*, *2*). The WWCR reduction approach discussed in this paper improves on previous modeling efforts by using an optimization algorithm to determine which specific interchanges should be considered for installation of advanced technology WWD countermeasures. This can save agencies money compared with equipping entire high-risk segments or corridors and could be used to prioritize investment as part of a long-term plan to install advanced technology WWD countermeasures at all ramps. The optimization algorithm can be modified to represent various real-world scenarios and provide a customized installation plan for any agency nationwide. It is very robust and can be used for networks ranging from single roadways to statewide systems. The approach could also be expanded to examine multiple types of WWD countermeasures and include all types of divided highways.

Literature Review

The first major research on WWD crashes was performed by the California Department of Transportation in the 1950s and focused on freeway median crossover crashes (*3*). Additional research in California in the late 1960s analyzed WWD events detected by cameras and road tubes and examined drivers who got WWD traffic citations to determine where WWD frequently occurred and potential explanatory characteristics (*4*).

More recently, other states have examined WWD crashes. North Carolina examined statewide freeway WWD crashes from 2000 to 2005 and found that they were only 0.2% of all freeway crashes, but resulted in 5.6% of freeway fatalities (*5*). Most freeway WWD crashes occurred during daytime hours with intoxicated and elderly drivers over-represented (*5*). Texas A&M Transportation Institute (TTI) has been studying WWD since 2003. They examined WWD crashes on Texas freeways from 1997 to 2000 and found similar results to North Carolina for WWD severity and the over-representation of intoxicated and elderly drivers. However, they also found that drivers younger than 35 were over-represented and that most WWD crashes occurred during nighttime hours (*6*, *7*). In Illinois, freeway WWD crashes from 2004 to 2009 were analyzed; similar results to Texas were found (*8*, *9*).

Since only 3% of crashes on high-speed divided highways are WWD crashes (*10*), it can be beneficial to examine other non-crash WWD events. TTI examined WWD traffic management center (TMC) logs and 911 calls around San Antonio starting in 2011 to determine corridors with high frequencies of WWD events (*11*). The University of Central Florida (UCF) has also been investigating non-crash WWD events since 2012. UCF's research started by examining WWD crashes, citations, 911 calls, TMC logs, and driver experiences in Central Florida (*12*, *13*). It then expanded statewide and modeled non-crash WWD events to predict the number of WWD crashes on Florida limited-access facilities (*14*, *15*). This modeling approach was then used in South Florida and Central Florida (the two main WWD hotspots in Florida) to predict the number of WWD crashes in roadway segments based on non-crash WWD events, interchange design, and traffic volumes (*1*, *2*).

The other main area of WWD research has focused on WWD countermeasures. Significant research has been done on enhancements to traditional countermeasures ("Wrong Way" signs and pavement markings) and advanced technology WWD countermeasures. Enhancements to traditional countermeasures include lowered sign mounting heights, larger signs, overhead signs, additional signs, reflective tape on signposts, and reflective pavement markers (*6*, *7*, *11*, *16*). There are also a variety of advanced technology WWD countermeasures; two examples include "Wrong Way" signs with light-emitting diodes (LEDs) around the border and "Wrong Way" signs equipped with RFBs. The LED signs have been used in many states, including Florida, Texas, and Wisconsin (*11*, *12*, *17*, *18*). However, TTI found that LED signs could be more difficult for intoxicated drivers to read than traditional signs, possibly reducing the effectiveness of these signs for intoxicated drivers (*19*). UCF researchers first proposed using two sets of RFB "Wrong Way" signs as WWD countermeasures; they were installed in Central Florida starting in 2015, where they have successfully detected and alerted wrong-way drivers, causing them to turn around before entering the mainline (*12*, *13*, *17*).

The research discussed in this paper uses optimization techniques to identify the best locations to install advanced technology WWD countermeasures. Optimization has been used for other traffic operations applications, such as developing tactical plans for construction projects, deploying traffic sensors, and managing traffic through ramp metering and variable speed limits (*20–22*). However, no previous research has used optimization techniques for the placement of WWD countermeasures. This novel use of optimization will allow agencies to best utilize their available resources by installing advanced technology WWD countermeasures at the ramps where it would be most cost-effective.

Research Objective

The main objective of this paper was to develop a WWCR reduction approach to help agencies prioritize the investment of available resources for installing advanced technology WWD countermeasures. This approach consists of the previously developed WWCR modeling and a newly developed optimization algorithm that can identify the priority interchanges for countermeasure installation based on investment levels and other constraints. An example application of the WWCR reduction approach is provided to illustrate how it can be used by agencies to represent different potential real-world scenarios. The major benefits of this approach are to help agencies determine where to install advanced technology WWD countermeasures based on available resources and provide savings compared with the WWCR segment model while still achieving high WWCR reduction.

Research Methodology

The WWCR reduction approach has two main components: the WWCR segment model and the WWD countermeasure optimization algorithm. This section discusses the research methodologies for each of these components, followed by a hypothetical example to show how agencies can apply these methodologies.

WWCR Segment Modeling Methodology

The WWCR segment model predicts the number of WWD crashes in each segment of a limited-access facility based on non-crash WWD events, geometric design data, and traffic volumes. This modeling methodology was previously developed by UCF and is discussed briefly in this section; details can be found in (*1*). Roadway segments are used since it is often not known where a wrong-way driver enters the facility. Each segment (k) contains the same number of interchanges (n) and overlaps with adjacent segments. Figure 1 shows an example of this overlapping for State Road (SR) 408 with $n = 5$. The first segment ($k = 1$) contains interchanges 1 (SR91) through 5 (SR435) (Figure 1*a*), while the second segment ($k = 2$) contains interchanges 2 (SR50) through 6 (Pine Hills Road) (Figure 1*b*). Each successive segment moves one interchange along the roadway in this same manner.

To develop the WWD crash prediction model, data on WWD crashes, citations, and 911 calls for multiple years are needed. Overlap between and within these datasets should be removed. These WWD event data should then be allocated to the roadway segments where they occurred. Data on interchange design and traffic volumes for each segment should be collected, enabling a Poisson regression model to be developed to predict the number of WWD crashes in each segment. A Poisson distribution is used since WWD crashes are rare events with count values (0, 1, 2, etc.). The developed model will show which factors significantly influence the number of WWD crashes in a segment. Adding the predicted crash values from this model to the actual crash numbers results in a WWCR value for each segment, which considers the actual and potential WWD crashes on the segment and can be used to identify WWD hotspots. If non-crash WWD event data are not available, a model can be developed to predict WWD crashes using only geometric design and traffic volume data.

Development of the WWD Countermeasure Optimization Algorithm

The second component of the WWCR reduction approach is the optimization algorithm. The main purpose of this algorithm is to identify the optimal interchanges where advanced technology WWD countermeasures should be installed based on the WWCR modeling results and available resources. This algorithm analyzes each interchange to determine the total WWCR reduction that would occur for all segments containing the interchange by equipping the interchange with these countermeasures. It also considers the cost of equipping the interchange. The algorithm can be used to determine the interchanges to equip with WWD countermeasures that maximize the WWCR reduction for a given investment level or to determine the optimal investment level that will provide the most cost-effective WWCR reduction. The results of the algorithm can help agencies decide on investment amounts and locations to install these countermeasures, but the results should not be used as the sole basis for installation decisions.

Before developing the algorithm, a few assumptions are needed. First, the WWCR reduction of the WWD countermeasure needs to be estimated. This can be estimated by various methods, such as examining WWD detection data, analyzing WWD event data, or reviewing previous studies of the countermeasure. If α = percentage reduction in WWCR due to installation of the WWD countermeasure, then it is expected that equipping all

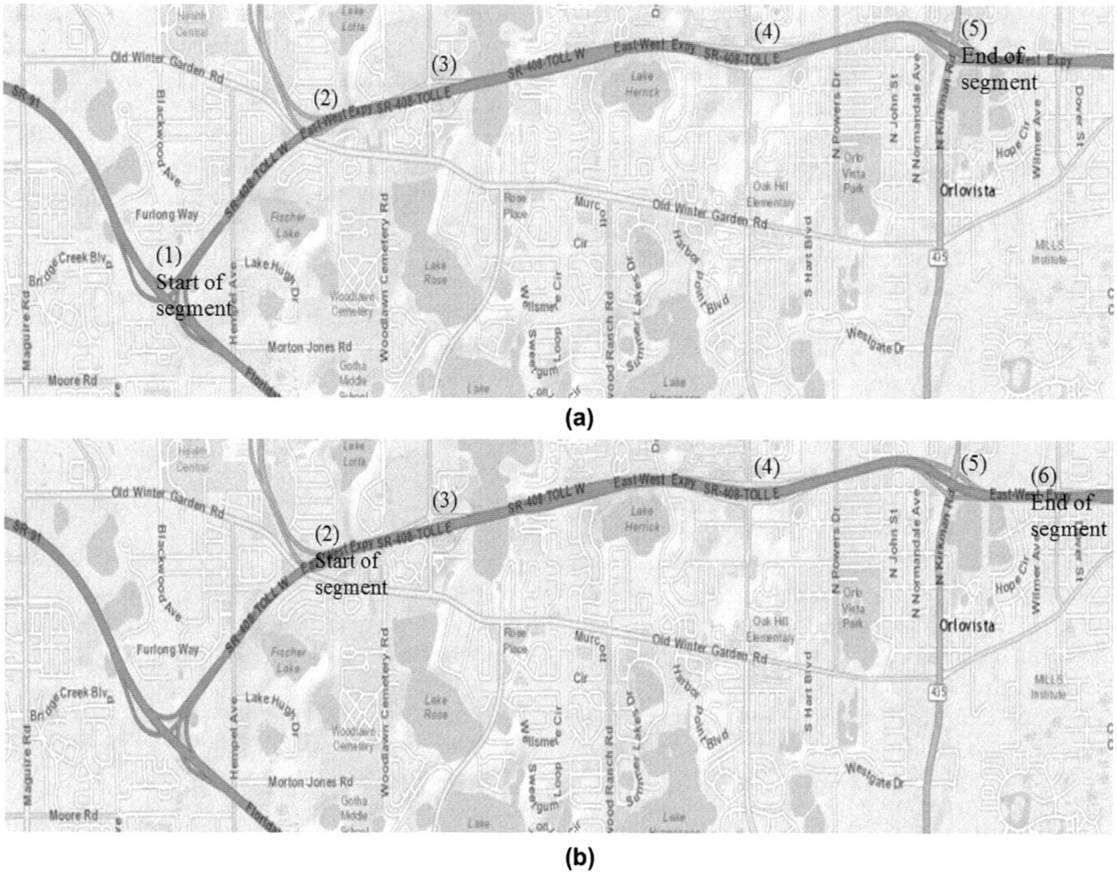

Figure 1. Example of roadway segmentation on SR 408 (maps created by UCF): (*a*) first five-interchange segment of SR 408 with interchange numbers and (*b*) second five-interchange segment of SR 408 with interchange numbers.

interchanges in a segment will reduce the segment's WWCR by α. It can also be assumed that all interchanges in a segment will equally contribute to the segment's WWCR reduction, as it is often difficult to know from which interchange the wrong-way driver entered. With this assumption, each interchange equipped with the advanced technology countermeasure reduces WWCR by α/n. For instance, if $\alpha = 50\%$ and $n = 5$, then each interchange can be assumed to reduce WWCR by $(50\%)/5 = 10\%$.

Another assumption regards the capital and installation cost of the advanced technology WWD countermeasure. While this cost will vary depending on the equipment purchased and the specific nature of each location, the estimated mean or median cost of installation can be used as a representative cost that is the same for all ramps. This means that the cost for each interchange will only depend on the number of exit ramps where a wrong-way vehicle could enter the mainline. Interchanges with multiple exit ramps will therefore cost more to equip than interchanges with one exit ramp. While it is possible to only equip the countermeasures on some exit ramps at an interchange, this was not considered as it is very difficult to determine exactly which exit ramp has the most WWD entries. Therefore, it was assumed that an interchange either has all or none of its exit ramps equipped with the countermeasure. Future improvements to the algorithm will look at each direction of a roadway separately so agencies can determine if only exit ramps in one direction should be equipped with countermeasures.

Formulation of Base Optimization Algorithm

Based on the discussed purpose and assumptions, a base optimization algorithm was developed that considers the WWCR reduction and available investment level. Additional constraints can be added to this base algorithm as desired; some of these constraints are discussed in the example application. The decision variable for this algorithm is a binary variable for each interchange:

$$X_{i,j} = \begin{Bmatrix} 1 \text{ if WWD countermeasure implemented at interchange } j \text{ on roadway } i, \\ 0 \text{ otherwise} \end{Bmatrix}, i \in I, j \in J_i$$

Where I = set of roadways being examined
J_i = set of interchanges on roadway i

The objective of this algorithm is to select the interchanges that maximize the WWCR reduction over the entire system being examined (Equation 1).

$$\text{Maximize } z = \sum_i \sum_k R_{i,k}, i \in I, k \in K_i \qquad (1)$$

Where z = objective function value
$R_{i,k}$ = WWCR reduction on segment k of roadway i due to installation of WWD countermeasures
K_i = set of segments on roadway i
The value of $R_{i,k}$ for each segment is determined by Equation 2:

$$R_{i,k} = (\alpha/_n)(WWCR_{i,k})\left(\sum_{j=k}^{j=k+n-1} X_{i,j}\right) \qquad (2)$$

Where $WWCR_{i,k}$ = WWD crash risk for segment k of roadway i before installing WWD countermeasures

The summation bounds $j = k$ and $j = k + n - 1$ ensure that only the interchanges located in segment k are considered. For instance, if $n = 5$ and $k = 2$, then the summation bounds are $j = 2$ and $j = 2 + 5 - 1 = 6$, indicating that segment 2 on roadway i contains interchanges 2 through 6.

The last component of the base optimization algorithm is the investment constraint. Since the cost of the countermeasure was assumed to be constant for each ramp, the total cost at the selected interchanges is:

$$\sum_i \sum_j N_{i,j} X_{i,j}, i \in I, j \in J_i \qquad (3)$$

Where $N_{i,j}$ = number of exit ramps at interchange j on roadway i

If B = maximum number of exit ramps where advanced technology WWD countermeasures can be installed based on investment level, then the investment constraint can be realized by setting Equation 3 to be less than or equal to B.

Example Application of the WWCR Reduction Approach

To demonstrate how the developed WWCR reduction approach can be used by agencies, a hypothetical example application is discussed. This example considers a two-phase installation strategy, showing how the approach can be used to prioritize investment as more resources become available. Various constraints are considered to show the flexibility of the optimization algorithm; these constraints can be adapted to any roadway network with minimal changes. There are other practical constraints (such as ease of construction, availability of power source, etc.) that were not considered which could influence real-world installation decisions. However, this algorithm can still help agencies decide where to install advanced technology WWD countermeasures.

The roadway network considered in this example is the Central Florida Expressway Authority (CFX) toll road system. CFX operates 130 exit ramps on five toll roads in Central Florida: SR408, SR414, SR417, SR429, and SR528. The advanced technology WWD countermeasure considered is two sets of "Wrong Way" signs equipped with RFBs. CFX was the first agency in the US to install two sets of these devices as WWD countermeasures in February 2015 and currently has them installed at 35 ramps on their system. As of June 2017, there have been 160 WWD detections by these RFB devices (excluding detections caused by emergency vehicles, maintenance vehicles, lawn mowers, or other similar vehicles), with 123 of the detected vehicles turning around. Using these values, $\alpha = (123/160) \times 100\% = 77\%$ for this example.

The example also uses the WWCR values obtained from the WWCR modeling application discussed in (*1*), which used five-interchange segments ($n = 5$). Therefore, it was assumed that each interchange equipped with RFBs reduces WWCR by (77%)/5 = 15.4%. Table 1 shows the starting and ending interchanges (with interchange numbers in parentheses) and WWCR values for the five highest WWCR segments that contain CFX interchanges on each CFX roadway (except for SR414, which only has one segment). The total WWCR for all segments that contain CFX interchanges is 63.78.

Example Application: Phase One

With these identified parameters, the developed optimization algorithm was used to optimally solve a hypothetical two-phase installation scenario. This scenario (both phases) and its constraints, parameters, and results are for demonstration purposes only and were not used by CFX to determine where to install RFBs on its system. In the first phase, CFX wants to determine the optimal investment level (number of ramps to equip with RFBs) to maximize the WWCR reduction on their system for minimal cost. It is assumed that CFX has no existing RFBs on their system and that they can obtain enough

Table 1. Segments with Highest WWCR on Each CFX Roadway

Roadway	Starting interchange (interchange number)	Ending interchange (interchange number)	WWCR value
SR408	Semoran Boulevard (16)	Rouse Road (20)	2.94
SR408	South Conway Road (15)	Dean Road (19)	2.90
SR408	Bumby Avenue (14)	SR417 (18)	2.87
SR408	Mills Avenue (13)	Goldenrod Road (17)	2.61
SR408	Orange Avenue (12)	Semoran Boulevard (16)	1.35
SR414	Orange Blossom Trail (1)	SR429 (4)	0.20
SR417	University Boulevard (19)	Lake Mary Boulevard (23)	3.87
SR417	SR50 (18)	Sanford Oveido Road (22)	2.73
SR417	SR408 (17)	Red Bug Lake Road (21)	2.19
SR417	Curry Ford Road (16)	Aloma Avenue (20)	1.70
SR417	Lee Vista Boulevard (15)	University Boulevard (19)	1.58
SR429	SR50 (9)	SR414 (13)	0.47
SR429	Plant Street (10)	Orange Blossom Trail (14)	0.42
SR429	US 192 (3)	Daniels Road (7)	0.25
SR429	SR91 (8)	Ocoee – Apopka Road (12)	0.25
SR429	Daniels Road (7)	West Road (11)	0.20
SR528	John Young Parkway (4)	Semoran Boulevard (8)	2.32
SR528	Orangewood Boulevard (3)	Tradeport Drive (7)	2.16
SR528	SR91 (5)	Goldenrod Road (9)	2.12
SR528	Sand Lake Road (6)	Narcoossee Road (10)	1.82
SR528	Tradeport Drive (7)	SR417 (11)	1.23

resources to equip up to 40 ramps. Additionally, CFX would like to equip at least ten interchanges with RFBs.

For this first phase, the optimization algorithm is used to determine the optimal number of ramps, B, that should be equipped with RFBs to provide the most cost-effective WWCR reduction. This optimal value of B can be determined by subtracting the percentage of ramps equipped from the percentage reduction in WWCR; this difference represents the percentage gain. The value of B with the largest percentage gain is the optimal number of ramps that should be equipped with RFBs. For this optimal value of B, the algorithm will identify the interchanges where RFBs should be equipped.

Constraints. To accurately represent this hypothetical scenario, additional constraints need to be added to the base optimization algorithm. The constraint in Equation 4 ensures that only CFX interchanges on the five studied Central Florida roadways are considered for RFB installation:

$$\sum_i \sum_j X_{i,j} = 0, \forall j \notin C \quad (4)$$

Where C = set of CFX interchanges

CFX also wants to equip at least 10 interchanges with RFBs, which can be represented by the following constraint:

$$\sum_i \sum_j X_{i,j} \geq 10, \forall j \in C \quad (5)$$

The last constraint is on the possible values of B. Since resources are available to equip up to 40 ramps, the maximum value of B is 40. Equation 5 requires that at least 10 interchanges are selected, so B cannot be less than 10. Since there are only nine CFX interchanges with one exit ramp, at least one interchange with two exit ramps must be equipped. Therefore, the minimum number of ramps that can be equipped is 11 (nine interchanges with one ramp plus one interchange with two ramps). So, B can be any integer value from 11 to 40.

Results. With these additional constraints, GAMS (General Algebraic Modeling System) was used to run the optimization algorithm for values of B from 11 to 40 (*23*). Table 2 shows the results for each value of B, including the percentage of CFX exit ramps equipped, WWCR reduction, the number of CFX interchanges equipped with RFBs, and the percentage gain. The percentage of CFX exit ramps equipped was calculated by dividing B by 130 (the total number of CFX exit ramps). The WWCR reduction percentage was calculated by dividing the WWCR reduction value by 63.78 (the total WWCR for all segments with CFX interchanges). The percentage gain is the difference between the WWCR reduction percentage and the percentage of CFX exit ramps equipped.

Table 2 shows that the maximum percentage gained (8.22%) occurs when B = 18 (shown in bold in Table 2), indicating that 18 ramps should be equipped with RFBs to achieve the most cost-effective WWCR reduction.

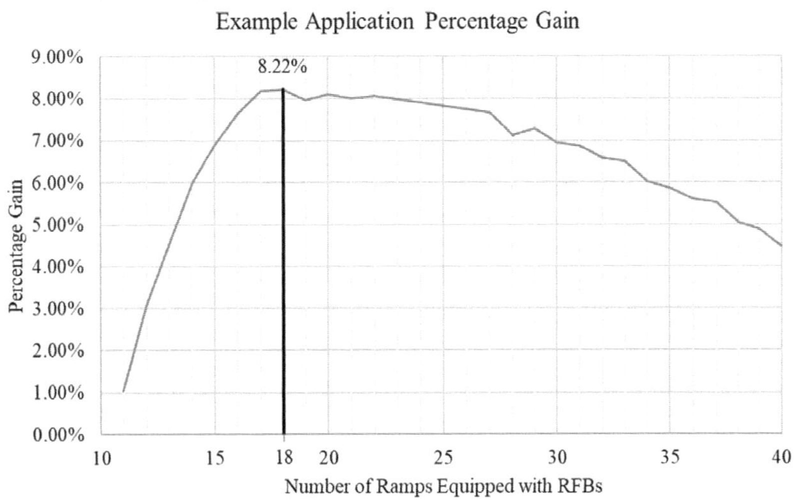

Figure 2. Percentage gain for first phase of example application.

This results in the following 10 interchanges being equipped with RFBs:

- SR408 interchange 2 at SR50 (one ramp)
- SR408 interchange 14 at Bumby Avenue
- SR408 interchange 15 at South Conway Road (one ramp)
- SR408 interchange 16 at Semoran Boulevard
- SR408 interchange 17 at Goldenrod Road
- SR408 interchange 19 at Dean Road
- SR417 interchange 16 at Curry Ford Road
- SR417 interchange 18 at SR50
- SR528 interchange 7 at Tradeport Drive
- SR528 interchange 9 at Goldenrod Road

All these interchanges have two exit ramps unless otherwise noted.

To better understand how the percentage gain varies as B increases, Figure 2 shows a graph of the percentage gain values from Table 2, with the black vertical line indicating the maximum percentage gain of 8.22% when $B = 18$. This graph shows that the percentage gain increases from $B = 11$ to $B = 17$, then remains relatively constant between 7.96% and 8.22% until $B = 23$, when it starts to steadily decrease until $B = 40$. The small variation in percentage gain from $B = 17$ to $B = 23$ indicates that any number of ramps between 17 and 23 could be equipped without significant decreases in overall effectiveness, but equipping 18 ramps is the most efficient use of resources.

The interchanges selected by the optimization algorithm were mainly in the high-WWCR segments shown in Table 1, indicating that the algorithm generally agrees with the WWCR model. However, the algorithm did not select all five interchanges in any one segment. This suggests that equipping all interchanges in the high-WWCR segments at once is not necessarily the most cost-effective way for agencies to spend their resources. To determine the potential savings provided by the optimization algorithm, the results from the algorithm and the WWCR segment model were compared (Table 3). The first two columns of Table 3 show the 10 interchanges that were selected by the optimization algorithm. The next two columns show the starting and ending interchange of the segment with the highest WWCR that contains each of the 10 selected interchanges. For instance, SR408 interchanges 16, 17, and 19 are all located in the segment from interchange 16–20, which has the highest WWCR on SR408 (Table 1). The last two columns show the CFX interchanges and number of ramps that are in the noted segments, but were not selected by the optimization algorithm. These represent locations where RFBs would be installed based on the WWCR segment model, but which were not selected by the optimization algorithm since they are not the most cost-effective locations.

Overall, Table 3 shows that only following the WWCR segment model would result in 24 more ramps being equipped with RFBs than the optimization algorithm suggests. The savings provided by the optimization algorithm can be calculated using Equation 6:

$$Savings = \left[{}^{N_M}/_{(N_O + N_M)}\right] x100\% \qquad (6)$$

Where N_O = number of ramps selected by optimization algorithm

N_M = number of additional ramps to equip from segment model

Table 2. Optimization Results of Example Application Phase One

B	CFX exit ramps equipped	WWCR reduction value	WWCR reduction	Number of CFX interchanges equipped	Gain
11	8.46%	6.06	9.51%	10	1.05%
12	9.23%	7.85	12.31%	10	3.08%
13	10.00%	9.29	14.56%	10	4.56%
14	10.77%	10.68	16.75%	10	5.98%
15	11.54%	11.73	18.40%	10	6.86%
16	12.31%	12.72	19.94%	10	7.63%
17	13.08%	13.56	21.26%	10	8.18%
18	**13.85%**	**14.07**	**22.07%**	**10**	**8.22%**
19	14.62%	14.40	22.58%	10	7.96%
20	15.38%	14.98	23.48%	10	8.10%
21	16.15%	15.40	24.15%	12	8.00%
22	16.92%	15.93	24.98%	11	8.06%
23	17.69%	16.37	25.67%	12	7.98%
24	18.46%	16.82	26.37%	12	7.91%
25	19.23%	17.26	27.06%	13	7.83%
26	20.00%	17.69	27.74%	13	7.74%
27	20.77%	18.13	28.43%	14	7.66%
28	21.54%	18.28	28.66%	13	7.12%
29	22.31%	18.87	29.59%	15	7.28%
30	23.08%	19.15	30.03%	14	6.95%
31	23.85%	19.60	30.72%	15	6.87%
32	24.62%	19.89	31.19%	15	6.57%
33	25.38%	20.34	31.88%	16	6.50%
34	26.15%	20.51	32.17%	17	6.02%
35	26.92%	20.91	32.78%	17	5.86%
36	27.69%	21.23	33.29%	16	5.60%
37	28.46%	21.67	33.98%	17	5.52%
38	29.23%	21.85	34.27%	18	5.04%
39	30.00%	22.25	34.88%	18	4.88%
40	30.77%	22.47	35.24%	17	4.47%

Note: Bold indicates that the maximum percentage gained (8.22%) occurs when $B = 18$.

For this example, $N_O = 18$ and $N_M = 24$, so savings $= [24/(18 + 24)] \times 100\% = 57\%$. This shows that the optimization algorithm can provide significant savings to agencies compared with the WWCR segment model alone by identifying which interchanges in the high-WWCR segments will provide the most WWCR reduction for the lowest cost. The optimization algorithm's savings are even greater when compared with a blanket installation approach on the three selected roadways (SR408, SR417, and SR528). This blanket installation would require CFX to equip 103 ramps with RFBs, resulting in a savings of $[(103–18)/103] \times 100\% = 83\%$ due to the optimization algorithm. These results show how the optimization algorithm can help agencies better utilize their available resources by only equipping high-risk interchanges instead of equipping entire segments or corridors at once.

Example Application: Phase Two

In the second phase of this hypothetical example, it is assumed that CFX has equipped the 18 ramps selected in the first phase and has obtained resources to equip RFBs at 20 additional ramps on their system. To provide better geographical coverage, they want to equip at least one interchange on each of their five roadways in this second phase. The optimization algorithm will be used to identify the interchanges where RFBs should be installed to maximize the WWCR reduction and meet the mentioned constraints, showing how agencies can use additional constraints to personalize the optimization algorithm for their network.

Constraints. This hypothetical second-phase installation scenario requires additional constraints to the base optimization algorithm that were not considered in the first phase. One common constraint to both phases is the one shown in Equation 4, which ensures that only CFX interchanges are considered. Since 10 interchanges are already equipped from the first phase, the following constraint is needed:

$$X_{ij} = 1, \forall i, j \in P \tag{7}$$

Where P = set of 10 interchanges equipped with RFBs in phase one

Table 3. Comparison between WWCR Model and Optimization Algorithm

Roadway	Interchanges selected by optimization (interchange number)	Highest WWCR segment starting interchange	Highest WWCR segment ending interchange	Additional CFX Interchanges in segment that were not selected by optimization	Number of additional ramps to equip from segment model
SR408	Semoran Boulevard (16)	Semoran Boulevard (16)	Rouse Road (20)	SR417 (18), Rouse Road (20)	6
	Goldenrod Road (17)	Semoran Boulevard (16)	Rouse Road (20)		
	Dean Road (19)	Semoran Boulevard (16)	Rouse Road (20)		
	South Conway Road (15)	South Conway Road (15)	Dean Road (19)	None	0
	Bumby Avenue (14)	Bumby Avenue (14)	SR417 (18)	None	0
	SR50 (2)	SR91 (1)	Kirkman Road (5)	Good Homes Road (3), Hiawassee Road (4), Kirkman Road (5)*	6
SR417	SR50 (18)	SR50 (18)	Sanford Oviedo Road (22)	University Boulevard (19)*	4
	Curry Ford Road (16)	Curry Ford Road (16)	Aloma Avenue (20)	SR408 (17)*	4
SR528	Tradeport Drive (7)	John Young Parkway (4)	Semoran Boulevard (8)	Semoran Boulevard (8)*	4
	Goldenrod Road (9)	SR91 (5)	Goldenrod Road (9)	None*	0
Total number of additional ramps to equip from segment model:					24

*SR408 interchange 1; SR417 interchanges 20, 21, and 22; and SR528 interchanges 4, 5, and 6 are operated by Florida's Turnpike Enterprise, not CFX, so they were not considered as additional CFX interchanges.

The second-phase scenario also requires that at least one interchange on each of the five roadways is equipped with RFBs. This can be represented by the following constraint:

$$\sum_{j} X_{i,j} \geq A_i + 1, \forall i \tag{8}$$

Where A_i = number of interchanges on roadway i equipped with RFBs in phase one
$A_i = 6$ for SR408, $A_i = 2$ for SR417, $A_i = 2$ for SR528, and $A_i = 0$ for SR414 and SR429.

Results. After adding these new constraints to the base optimization algorithm, it was run in GAMS for a total of 38 ramps (18 ramps selected in the first phase plus the 20 new ramps being added in the second phase). The following eight interchanges were selected for this second phase:

- SR408 at Good Homes Road (two ramps)
- SR408 at SR417 (four ramps)
- SR414 at Keene Road (one ramp)
- SR417 at Lee Vista Boulevard (two ramps)
- SR417 at University Boulevard (four ramps)
- SR429 at Orange Blossom Trail (one ramp)
- SR528 at Semoran Boulevard (four ramps)
- SR528 at Narcoossee Road (two ramps)

Since SR414 and SR429 had low WWCR compared with the other three roadways, the algorithm selected the cheapest interchanges to equip on these roadways, leaving more resources for the roadways with higher WWCR.

Summary and Conclusion

WWD on high-speed limited-access facilities often results in head-on collisions with severe injuries or fatalities. Most WWD crashes occur at night and are caused by intoxicated drivers, so traditional "Wrong Way" signs and pavement markings might not be sufficient to prevent these wrong-way drivers from entering the mainline. To alert these unintentional or confused wrong-way drivers, advanced technology WWD countermeasures, such as "Wrong Way" signs equipped with RFBs, could be needed. It can be expensive for agencies to install these WWD countermeasures on their roadway network, so a method to determine the most cost-effective locations for these countermeasures based on WWD event history, geometric design, traffic volumes, and countermeasure cost is needed. No previous research has developed such a method.

This paper's research has considered all the factors mentioned above to help agencies determine the optimal locations to install advanced technology WWD countermeasures. A WWCR reduction approach consisting of a WWCR segment model and optimization algorithm was developed that can maximize the WWCR reduction on any limited-access roadway network for a given investment level. In this approach, a WWCR model is developed that considers WWD events (crashes and noncrashes), geometric design of interchanges, and traffic volumes to determine the WWCR on overlapping multi-interchange segments of limited-access facilities (*1*).

Using these segment WWCR values, the developed optimization algorithm can help agencies identify priority interchanges to consider installing WWD countermeasures at based on investment level and other practical constraints. The algorithm can also be used to determine the optimal investment level to provide the most cost-effective WWCR reduction. This WWCR reduction approach can help agencies better utilize their available resources and, if desired, be used as part of a long-term plan to install these countermeasures at all exit ramps.

A hypothetical example application demonstrated how the WWCR reduction approach can be used by agencies. In this example, the optimal locations to install "Wrong Way" signs with RFBs on the CFX toll road system using a two-phase installation strategy were determined. For the first phase, the optimal investment level that provided the most cost-effective WWCR reduction while ensuring that at least 10 interchanges were equipped with RFBs was determined. At optimality, the algorithm selected 18 ramps at 10 interchanges for RFB installation, providing a gain of 8.22%. These interchanges were located on three different CFX roadways and no five-interchange segment had all its interchanges selected. The optimization algorithm provided savings of 57% compared with equipping entire high-WWCR segments and 83% compared with equipping all interchanges on the selected roadways. These savings indicate that equipping all interchanges in the high-WWCR segments is not the most cost-effective way to combat WWD. Without this newly invented optimization algorithm, agencies would not recognize these potential savings, indicating the significant value this algorithm has added to the segment model developed in (*1*).

For the second phase, the optimal locations to install RFBs for a set investment level of 20 ramps were determined. Constraints were added to ensure that the interchanges chosen in the first phase were equipped and at least one additional interchange on each CFX roadway was selected. The optimization algorithm selected eight interchanges for RFB installation in this second phase. This example shows how additional constraints concerning roadway coverage and existing WWD countermeasures can be incorporated into the optimization algorithm to represent different real-world scenarios.

Overall, the developed WWCR reduction approach can be very beneficial to roadway agencies. It can be used to determine the optimal investment level to spend on WWD countermeasures or help agencies determine the best locations at which to install these countermeasures for a set investment level, ensuring that resources are not wasted by installing them at locations where they are not needed right away. By revisiting the approach whenever additional funds to combat WWD are available, agencies can obtain priority lists of locations to equip with countermeasures. Furthermore, the optimization algorithm can be adapted to an agency's specific requirements by incorporating additional (or different) constraints, such as the ones shown in the example application. This WWCR reduction approach is very robust and scalable, as it can be adapted to any type of limited-access facility network, from statewide systems to networks within one county or even single roadways. It can also be transferred and applied to other networks in the country or expanded to cover all divided highways.

Further research will be conducted to improve the capabilities of the developed WWCR reduction approach. These improvements will include relaxing some of the assumptions to make the optimization algorithm more representative of the real world. Improvements will also be made to the WWCR model by considering the direction of WWD events to allow for identification of individual ramps where countermeasures should be installed. Comparison between different WWD countermeasures and enhancements to traditional WWD countermeasures will also be considered, allowing agencies to determine the optimal placement of multiple types of WWD countermeasures simultaneously. With these improvements, the WWCR reduction approach's benefits shown in this paper will be increased, allowing agencies to efficiently utilize their resources while effectively reducing WWD. The approach can also be utilized by agencies with limited resources, allowing them to install advanced technology WWD countermeasures at a few interchanges instead of having to install them on entire corridors. This WWCR reduction approach can help agencies improve traffic operations and safety, while providing them with savings that can be used to improve operational or other aspects of travel.

Acknowledgments

The research discussed in this paper was funded by the Central Florida Expressway Authority.

Author Contributions

The authors confirm contribution to the paper as follows: study conception and design: Sandt, Al-Deek; data collection: Sandt, Al-Deek; analysis and interpretation of results: Sandt, Al-Deek; draft manuscript preparation: Sandt, Al-Deek. All authors reviewed the results and approved the final version of the manuscript.

References

1. Sandt, A., H. Al-Deek, and J. Rogers. Identifying Wrong-Way Driving Hotspots by Modeling Crash Risk and Assessing Durations of Wrong-Way Driving Events. *Transportation Research Record: Journal of the Transportation Research Board*, 2017. 2616: 58–68.

2. Rogers, J., H. Al-Deek, A. Alomari, E. Gordin, and G. Carrick. Modeling the Risk of Wrong Way Driving on Freeways and Toll Roads. *Transportation Research Record: Journal of the Transportation Research Board*, 2016. 2554: 166–176.
3. Sicking, D., and K. Lechtenberg. Guidelines for Implementation of Cable Median Barrier. *Transportation Research Record: Journal of the Transportation Research Board*, 2009. 2120: 82–90.
4. Tamburri, T. *Wrong-Way Driving Accidents Are Reduced.* Highway Research Record #292. California Division of Highways, 1969.
5. Braam, A. C. *Wrong Way Crashes: Statewide Study of Wrong Way Crashes on Freeways in North Carolina.* Traffic Engineering and Safety Systems Branch, North Carolina Department of Transportation, 2006.
6. Cooner, S. A., A. S. Cothron, and S. E. Ranft. *Countermeasures for Wrong-Way Movement on Freeways: Overview of Project Activities and Findings.* Report No. FHWA/TX-04/4128-1. Texas Transportation Institute, College Station, 2004.
7. Cooner, S. A., and S. E. Ranft. Wrong-Way Driving on Freeways: Problems, Issues, and Countermeasures. Presented at 87th Annual Meeting of the Transportation Research Board, Washington, D.C., 2008.
8. Zhou, H., J. Zhao, M. Pour-Rouholamin, and P. Tobias. Statistical Characteristics of Wrong-Way Driving on Illinois Freeways. *Traffic Injury Prevention*, Vol. 16, No. 8, 2015, pp. 760–767. http://dx.doi.org.ezproxy.net.ucf.edu/10.1080/15389588.2015.1020421.
9. Zhou, H., J. Zhao, R. Fries, M. R. Gahrooei, L. Wang, B. Vaughn, K. Bahaaldin, and B. Ayyalasomayajula. *Investigation of Contributing Factors Regarding Wrong-Way Driving on Freeways.* Publication FHWA-ICT-12-010. Illinois Center for Transportation, Urbana, 2012.
10. National Transportation Safety Board. *Wrong-Way Driving.* Highway Special Investigation Report NTSB/SIR-12/01, Washington, D.C., 2012.
11. Finley, M., S. Venglar, V. Iragavarapu, J. Miles, E. Park, S. Cooner, and S. Ranft. *Assessment of the Effectiveness of Wrong Way Driving Countermeasures and Mitigation Methods.* Report 0-6769-1. Texas A&M Transportation Institute, College Station, 2014.
12. Sandt, A., H. Al-Deek, J. Rogers, and A. Alomari. Wrong-Way Driving Prevention: Incident Survey Results and Planned Countermeasure Implementation in Florida. *Transportation Research Record: Journal of the Transportation Research Board*, 2015. 2484: 99–109.
13. Rogers, J., H. Al-Deek, A. Alomari, F. Consoli, and A. Sandt. Wrong-Way Driving on Florida Toll Roads: An Investigation into Multiple Incident Parameters and Targeted Countermeasures for Reductions. *International Journal of Engineering Management and Economics (Special Issue on Transport)*, Vol. 5, No. 3/4, 2015, pp. 144–168.
14. Rogers, J., A. Sandt, H. Al-Deek, A. Alomari, N. Uddin, E. Gordin, C. Dos Santos, J. Renfrow, and G. Carrick. Wrong-Way Driving: Multifactor Risk-Based Model for Florida Interstates and Toll Facilities. *Transportation Research Record: Journal of the Transportation Research Board*, 2015. 2484: 119–128.
15. Rogers, J., A. Sandt, H. Al-Deek, and A. Alomari. Wrong-Way Driving Multifactor Prediction Models for Florida Limited Access Facilities. *Proc., Road Safety and Simulation International Conference 2015 RSS2015,* Compendium of Papers. Southeastern Transportation Center, Orlando, FL, October 2015.
16. Zhou, H., and M. Pour-Rouholamin. *Guidelines for Reducing Wrong-Way Crashes on Freeways.* Publication FHWA-ICT-14-010. Illinois Center for Transportation, Urbana, 2014.
17. Sandt, A., H. Al-Deek, J. Rogers, A. Alomari, and E. Gordin. Modeling Driver Responses to Wrong-Way Driving Countermeasures through a Driver Survey and Countermeasure Implementation in Florida. Presented at 95th Annual Meeting of the Transportation Research Board, Washington, D.C., 2016.
18. Wilson, B. Is There a Right Way? *Roads & Bridges*, 2013, Vol. 50, No. 1, pp. 40-42.
19. Finley, M., J. Miles, and E. S. Park. Closed-Course Study to Examine the Effect of Alcohol-Impairment on a Driver's Ability to Identify and Read Signs. *Transportation Research Record: Journal of the Transportation Research Board*, 2017. 2660: 86–93.
20. Faghih-Imani, A., and L. Amador-Jimenez. From Strategic Optimization to Tactical Plans: Coordinating Treatments on Road Infrastructure. Presented at 92nd Annual Meeting of the Transportation Research Board, Washington, D.C., 2013.
21. Li, X., and Y. Ouyang. Reliable Traffic Sensor Deployment Under Probabilistic Disruptions and Generalized Surveillance Effectiveness Measures. *Operations Research*, Vol. 60, No. 5, 2012, pp. 1183–1198. http://dx.doi.org/10.1287/opre.1120.1082.
22. Chow, A. Optimisation of Dynamic Motorway Traffic Via a Parsimonious and Decentralised Approach. *Transportation Research Part C: Emerging Technologies*, Vol. 55, 2015, pp. 69–84. http://dx.doi.org/10.1016/j.trc.2015.01.009.
23. Software Program. *General Algebraic Modeling System 24.8.3.* GAMS Development Corporation, Washington, D.C., 2017.

The Standing Committee on Freeway Operations (AHB20) peer-reviewed this paper (18-01089).

This paper and its contents, including conclusions and results, are solely those of the authors; they do not represent opinions or policies of the funding agency.

Article

TRR
JOURNAL OF THE TRANSPORTATION RESEARCH BOARD

Analysis of Performance Data Collected from Two Wrong-Way Driving Advanced Technology Countermeasures and Results of Countermeasures Stakeholder Surveys

Transportation Research Record
2018, Vol. 2672(14) 96–105

DOI: 10.1177/0361198118778933
journals.sagepub.com/home/trr
SSAGE

Md Imrul Kayes[1], Haitham Al-Deek[1], Adrian Sandt[1], John H. Rogers, Jr.[1], and Grady Carrick[2]

Abstract

Wrong-way driving (WWD) often leads to severe collisions that cause serious injuries and deaths. Conventional "Wrong Way" signs can reduce WWD events, but can be insufficient in some cases. In areas with many WWD events, transportation agencies can be proactive by considering the use of countermeasures with advanced technologies to actively warn motorists of WWD violations. This paper analyzes recent performance data collected from two types of advanced technology WWD countermeasures implemented in Florida: light-emitting diode (LED) signs in South Florida and rectangular flashing beacon (RFB) signs in Central Florida. The 17 LED sites experienced a 38% reduction in WWD citations and 911 calls after the signs were installed. Images taken by the on-site cameras were examined to see how many vehicles turned around for both the RFB and LED treatment sites. Over 77% of the 170-detected wrong-way vehicles self-corrected their wrong-way movement at the RFB sites (each with two sets of signs and multiple cameras) and 14% self-corrected at the LED sites (each with one set of signs and one camera). Surveys were also conducted regarding these two WWD countermeasures. More than 73% of the 2,052 respondents preferred RFBs over LEDs, mainly due to the double set of RFB signs and their flashing pattern. The performance and survey results show that both the LEDs and RFBs have effectively reduced WWD movements. However, modifications could be made to both countermeasures to improve their detection ability and make wrong-way drivers more likely to turn around.

Wrong-way driving (WWD) occurs when a driver intentionally or unintentionally drives opposite the direction of traffic flow. The potential risk of WWD crashes is a major concern, especially on high speed, limited-access facilities where the chances of serious injuries or death greatly increase. Florida is the third highest state in the nation for WWD fatalities (*1*). There were a total of 280 WWD crashes from 2009 to 2013 on Florida's freeways and expressways, which resulted in 400 injuries and 75 deaths (*2*).

Given the severity of this problem, it is important for transportation agencies to be proactive in preventing or reducing WWD events. Historically, different agencies throughout the United States have used signs and pavement markings to notify wrong-way drivers, but these signs and markings are often not sufficient, particularly for notifying impaired drivers. Alerting these drivers is important since previous studies have found that WWD is commonly caused by intoxicated drivers during nighttime hours (*3–7*). Static signs and pavement markings might also not have the same effects for older drivers or unfamiliar drivers in unknown areas, especially at night due to limited visibility and lighting.

WWD countermeasures that actively detect and warn wrong-way drivers have better potential to reduce these events, but these countermeasures still need to be thoroughly evaluated using real world data and analysis. Two Florida agencies have already implemented intelligent transportation system (ITS) devices that include "Wrong Way" signs with flashing beacons or light-emitting diodes (LEDs) to alert wrong-way drivers and increase the chance of them self-correcting their WWD behavior before causing a crash. The Florida Turnpike Enterprise (FTE) has installed "Wrong Way" signs illuminated with LEDs at 17 exit ramps on two limited-access facilities in South Florida: State Road (SR) 821 and SR 869. The Central Florida Expressway Authority (CFX) has installed "Wrong Way" signs with rectangular flashing beacons (RFBs) at 35 exit ramps on six

[1]Department of Civil, Environmental, and Construction Engineering, University of Central Florida (UCF), Orlando, FL
[2]Enforcement Engineering, Inc., Saint Johns, FL

Corresponding Author:
Address correspondence to Haitham Al-Deek: Haitham.Al-Deek@ucf.edu

limited-access facilities in Central Florida: SR 408, SR 414, SR 417, SR 429, SR 451, and SR 528.

This paper analyzes the latest data available from these two WWD countermeasures (through June 30, 2017, for LEDs and July 24, 2017, for RFBs). These data consist of WWD 911 dispatch calls, WWD citations, and WWD detection data. Before-and-after WWD citation and 911 call data were used to evaluate the effectiveness of LEDs. WWD crashes were not used for this statistical analysis because of the rarity of these events, as WWD crashes account for only 3% of freeway crashes (*7*). The noncrash WWD event data were used for this analysis since they represent events detected by law enforcement and have a sufficient sample for statistical reliability. These data sets were used to develop a WWD event (911 dispatch calls and citations) modification factor (similar to crash modification factor) via a yoked comparison method (*8*). This modification factor shows the percent change in WWD events due to the implementation of the LED signs.

For the RFB signs, a different evaluation method was required as there was insufficient data in the after period. The RFBs were evaluated by examining images captured by cameras mounted on the RFB signs at the treated ramps to see if detected wrong-way vehicles turned around. Calculating the percentage of wrong-way vehicles that turned around provides a clear indicator of RFB performance. The RFBs' turnaround percentage was compared with the LEDs' turnaround percentage to identify any differences between the two technologies.

In addition to these performance data, surveys were also conducted to seek stakeholders' and road users' opinions about these two WWD countermeasures. University of Central Florida (UCF) conducted three different surveys for Florida Highway Patrol (FHP) officers and Florida Department of Transportation (FDOT) officials, FTE customers and Florida interstate drivers, and CFX customers. Survey participants were asked to choose their preferred WWD countermeasure (LED or RFB "Wrong Way" signs) for alerting and stopping wrong-way drivers. They were also asked to provide reasons for their preferences. Using the information from these surveys could help agencies improve their implemented countermeasures to make them more effective.

Literature Review

Sign and pavement markings have been used by transportation agencies to combat WWD for many years. In 1967, California started using road tubes and cameras to detect wrong-way drivers and better understand where WWD events tended to occur (*9*). In 1973, they lowered the height of "Do Not Enter" and "Wrong Way" signs and mounted them together on the same post. These changes decreased the rate of WWD per month from 50–60 to 2–6 in most of the problematic areas (*10*). In 1988, Campbell and Middlebrooks studied the effect of WWD countermeasures that included lowering "Wrong Way" signs, placing a stop bar at the end of the exit ramp, and installing yellow ceramic buttons at an exit ramp in Atlanta, Georgia. Their study concluded that more than 97% of WWD maneuvers were reduced within a month (*11*). To find the effectiveness of lowering the WWD signs, the North Texas Tollway Authority experimented with different configurations of lowered (2-ft above the ground) "Wrong Way" and "Do Not Enter" signs at 28 of 142 exit ramps on their system. The experiments suggested that equipping entire freeways would be needed to better determine whether these signs were more effective than traditional signage (*12*).

Pavement markings, such as red reflective raised pavement markers, pavement arrows for wrong way, and colored edge lines on exit ramps, are some of the widely used WWD countermeasures suggested by the *Manual on Uniform Traffic Control Devices* (*13*). Raised pavement markers were installed on Virginia and Texas off-ramps to alert wrong-way drivers (*14*) and the Virginia Highway and Transportation Research Council conducted a test on 16 people to evaluate these markers. Test subjects said the marking system was effective and 94% thought these markers would lead to corrective action by the drivers (*15*). Although these signs and pavement markers are a good way to notify wrong-way drivers, they are still ineffective for nighttime impaired drivers. Texas A&M Transportation Institute (TTI) researchers conducted a test to see the impact of these traditional WWD signs and pavement markings on alcohol-impaired drivers; they found that alcohol-impaired drivers look less to the left and right and more toward the front of the vehicle on the ground (*16*). However, lowering the height of the sign did not aid the alcohol-impaired drivers in identifying these signs (*16*). Measures like enlarging the sign, adding red retroreflective tape on the post, and incorporating red flashing LEDs around the sign could increase the attention of intoxicated drivers. However, intoxicated drivers had to be closer to LED signs than regular signs to be able to read them clearly (*12, 16*).

Another way to alert impaired wrong-way drivers is the use of ITS technologies. These technologies include cameras, radar, or circle locators to alert both wrong-way drivers and right-way drivers simultaneously. Existing ITS infrastructures, such as dynamic message signs and LED signs can be used to alert right-way drivers (*17*). New Mexico developed a directional traffic sensor system to warn detected wrong-way drivers on exit ramps (*13, 17*). The Washington State Department of Transportation used a system where LEDs, flashers, and video cameras were used to warn wrong-way drivers (*13, 17*).

In March 2007, the Harris County Toll Road Authority in Houston, Texas, installed a radar-based WWD detection system on the Westpark Toll Road. From 2008 to 2013 a total of 94 wrong-way vehicles self-corrected due to these countermeasures (*18*). A task force in San Antonio, Texas, evaluated the effectiveness of LED border-illuminated "Wrong Way" signs installed along U.S. Highway 281 and found a 35.4%

reduction in monthly 911 calls (*19*). FDOT installed a single set of RFBs at 16 interstate exit ramps in Tampa, Florida (*20*). Ozkul and Lin collected and analyzed six months of video data from two of these ramps; they found that 66 of 78 (85%) and 3 of 5 (60%) wrong-way drivers that triggered the radar at these two ramps self-corrected and turned around (*21*). Rapid rectangular flashing beacons (RRFBs), which have a different flashing pattern and frequency to the RFBs used as WWD countermeasures, have also been effective at getting motorists to yield at pedestrian installations in St. Petersburg, Florida (*22*). Caltrans and the California Highway Patrol have been conducting a pilot program to stop wrong-way crashes since January 2017. This program includes oversized "Do Not Enter" signs with red reflectors at 17 locations and solar-powered radar detector systems at six additional locations. Two wrong-way drivers have been detected since the beginning of this project, with one of these drivers causing a crash (*23*).

This previous research shows that WWD crashes are severe and continue to occur. Treatments with advanced technologies, in addition to traditional countermeasures at locations with high risk of WWD, can help prevent and reduce WWD activities. RFB and LED "Wrong Way" signs are two such advanced technology WWD countermeasures that are being used by transportation agencies in the United States. However, a comprehensive performance evaluation of these devices, especially double sets of RFBs, has not been conducted. This paper will evaluate for the first time, to the authors' knowledge, the performance of the double sets of RFBs that have been installed in Central Florida, as well as the LED signs installed in South Florida. More than two and a half years of data were collected and analyzed from these two countermeasures to evaluate their effectiveness using various methods.

Research Objectives

The main objectives of this paper are to determine the effectiveness of LEDs and RFBs in deterring WWD and how these countermeasures could be improved by incorporating aspects from each other. By observing and analyzing the data collected from 52 Florida locations where these two countermeasures are installed, as well as responses from surveys of transportation professionals and road users, the effectiveness of these treatments can be determined.

To achieve these objectives, this research utilized three different evaluation methods. First, before–after analyses of WWD events were performed on the LED sites in South Florida. Then, pictures of wrong-way vehicles from the RFBs and LEDs were examined to see how many vehicles turned around due to these devices. Finally, the responses of survey participants, including law enforcement officers, agency officials, and toll road and interstate travelers, were collected and examined to see their preferences and opinions about these two WWD countermeasures.

Description of LED and RFB Technologies

The two WWD countermeasures with advanced technologies being evaluated have different characteristics and field setups. Figure 1 shows an example setup for each of these devices. Each of the LED test sites has only one pair of "Wrong Way" signs (Figure 1*a*). These signs have front and rear radar to detect wrong-way vehicles. Once a vehicle is detected, the LED lights blink in a constant pattern and an alert is generated. Images of the wrong-way vehicle are also taken using a high-quality camera mounted on the signs so the traffic management center (TMC) can verify that there is a wrong-way vehicle. The front radar activates the sign and triggers the camera when a wrong-way vehicle enters the ramp. If the wrong-way vehicle continues and passes the rear radar, the TMC and FHP receive an audible alarm (*24*).

Unlike the LED sites, each ramp that is equipped with RFBs has four "Wrong Way" signs (two near the front of the ramp and two closer to the mainline). Each of these four "Wrong Way" signs has two sets of RFBs, one set on the top and one set on the bottom of the sign (Figure 1*b*). These RFBs flash in an alternating pattern when a wrong-way vehicle is detected. The first set of signs closer to the front of the ramp includes two cameras (oncoming and side) and detectors mounted with the RFBs. Two different detection technologies have been implemented at different exit ramps: radar and laser. The radar system consists of four detection zones that must be triggered by a wrong-way vehicle for a WWD alert to be sent to the TMC. The laser system consists of two detection zones: the first zone is positioned close to the intersection and triggers both sets of RFBs, while the second zone, which is positioned farther up the ramp, also triggers both sets of RFBs and activates the cameras.

LEDs Evaluation

Methodology

To evaluate the effectiveness of the LED "Wrong Way" signs, before–after analyses were conducted. LED signs were implemented on FTE roadways on October 2014 at 11 interchanges. All signs were installed during the month of October 2014, so this month was excluded from the analysis. The before-and-after periods were February 2012 to September 2014 (32 months) and November 2014 to June 2017 (32 months) respectively. As there were insufficient WWD crashes during the study period and the crash data for 2016 and 2017 were not available for analysis, only WWD 911 call and citation data were used for the evaluation of LED signs. UCF and TTI researchers found that WWD 911 calls and citations have a correlation with the number of WWD crashes (*3*, *19*, *25–29*). Three separate before–after analyses were conducted for three data groups: 911 calls, citations, and WWD events (911 calls and citations combined). A yoked comparison group study method was used to obtain WWD event modification factors for these data

Figure 1. WWD countermeasures with advanced technologies installed in Florida: (*a*) LED "Wrong Way" signs installed on FTE roadways (Courtesy of FTE), and (*b*) RFB "Wrong Way" signs installed on CFX roadways (Courtesy of CFX).

groups. This type of study "is a special case of the comparison group study where a single comparison site is matched to each treatment site based on similar geometric and traffic volume conditions" (*8*). This method was selected due to the small number of treated sites. The event modification factor obtained from the yoked comparison group study is a multiplicative factor used to calculate the expected number of WWD events after implementing a given countermeasure at a specific location.

For this yoked comparison, 11 interchanges without LEDs were selected as the comparison group. These interchanges were selected based on geometric characteristics and traffic volumes so they had similar characteristics to the 11 treated interchanges. The treatment group included six interchanges on SR 821 and five interchanges on SR 869; the comparison group included interchanges on SR 821, SR 869, and SR 91. Figure 2 shows the interchange locations for both the treatment and comparison groups. All interchanges in both the treatment and comparison groups were on toll roads located in Miami-Dade and Broward counties in South Florida.

Data Collection

Most WWD events are captured by public calls to law enforcement agencies and by officers who stop and cite drivers for WWD violations. Both WWD 911 and citation data

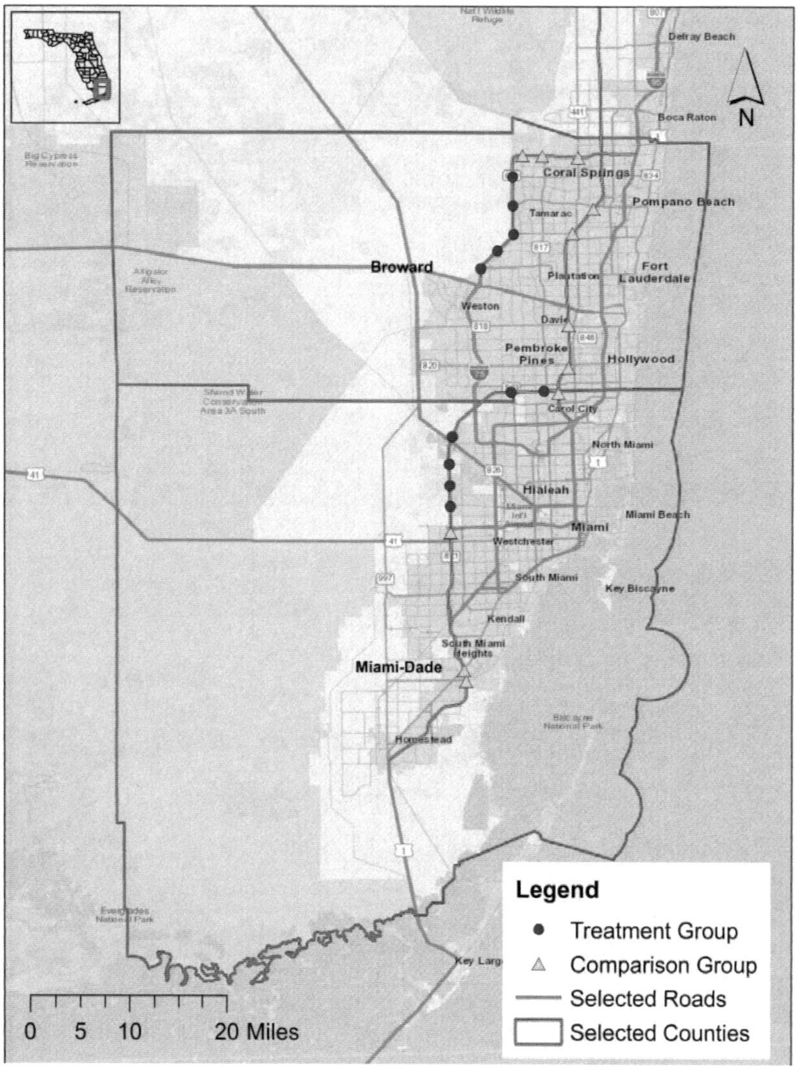

Figure 2. Treatment group and comparison group interchanges.

for the study period were provided by law enforcement. Global positioning system location information is a reliable part of 911 call and citation data in Florida. Each 911 call and WWD traffic citation are accompanied by longitude and latitude coordinates, which were plotted at FDOT interchange locations and the distance between them for the studied roadways (*30*). Google Maps software was used to verify the nearest interchanges from the longitudes and latitudes, and any overlap within and between citations and 911 calls was removed.

Before–After Study of LED Signs

Analysis of the collected data showed that the total number of WWD 911 calls and citations increased for both the treatment and comparison groups in the after period of the analysis, with the comparison group having a higher rate of increase. It is possible that these increases are due to the higher number of licensed drivers in the after period (*31*), which could result in higher traffic volumes. Also, there was a large WWD media campaign in South Florida that emphasized the serious effects of WWD. Consequently, people became more aware of wrong-way drivers and started reporting these incidents to law enforcement as soon as they saw them, which could have caused these increases in the after period.

The total number of WWD 911 calls and citations for the treatment and comparison groups are summarized in Table 1. The table shows the 32-month analysis duration, along with the results of the before–after analyses. Even though the

Table 1. Observed Counts and Results of Before–After Analyses

	Total counts at treatment group		Total counts at comparison group			
WWD	Before	After	Before	After	Modification factor	Standard error
911 Calls	30	34	18	36	0.51	0.174
Citations	7	13	9	16	0.79	0.378
Events	37	47	27	51	0.62	0.186

Note: WWD = wrong-way driving.

actual number of events increased in the after study, this increase would have been significantly greater if there was no treatment at the equipped sites. The results indicated that there was a 49% reduction in WWD 911 calls at a 99% confidence level and a 38% reduction in combined WWD events (911 calls and citations) at a 95% confidence level for the treatment sites due to the implementation of the LED signs. The results also showed a 21% reduction in WWD citations at the treatment sites, but this reduction was not statistically significant due to the smaller sample size of citations. The importance of the reduction in WWD events is that they could lead to a lower risk of potential WWD crashes at the treated sites.

The formulas that were used for the above analyses are given below (*8*). The expected number of WWD events for the treatment group that would have occurred in the after period without treatment ($N_{expected,T,A}$) is estimated from Equation 1.

$$N_{expected,T,A} = N_{observed,T,B} \times \left(\frac{N_{observed,C,A}}{N_{observed,C,B}} \right) \quad (1)$$

where

$N_{observed,T,B}$ = observed number of WWD events in the before period for the treatment group.

$N_{observed,T,A}$ = observed number of WWD events in the after period for the treatment group.

$N_{observed,C,B}$ = observed number of WWD events in the before period for the comparison group.

$N_{observed,C,A}$ = observed number of WWD events in the after period for the comparison group.

The variance of $N_{expected,T,A}$ is estimated by Equation 2. Equations 3 and 4 were used to determine the WWD event modification factor and the variance of this factor.

$$Var(N_{expected,T,A}) = N^2_{expected,T,A} \times \left(\frac{1}{N_{observed,T,B}} + \frac{1}{N_{observed,C,B}} + \frac{1}{N_{observed,C,A}} \right) \quad (2)$$

$$Event\ Modification\ Factor\ (EMF) = \frac{\dfrac{N_{observed,T,A}}{N_{expected,T,A}}}{1 + \dfrac{Var(N_{expected,T,A})}{N^2_{expected,T,A}}} \quad (3)$$

$$Variance(EMF) = \frac{EMF^2 \left[\dfrac{1}{N_{observed,T,A}} + \dfrac{Var(N_{expected,T,A})}{N^2_{expected,T,A}} \right]}{\left[1 + \dfrac{Var(N_{expected,T,A})}{N^2_{expected,T,A}} \right]^2} \quad (4)$$

RFB Evaluation

Methodology

CFX installed RFBs at 35 exit ramps on their roadway system starting in February 2015, with most of these installations occurring in the second half of 2016. These installations are the first locations in the United States to be equipped with two sets of RFB "Wrong Way" signs, as previous applications only used one sign. This innovation makes the CFX RFB installations unique and worthy of study. Since most of the treated ramps were equipped with RFBs very recently, a before–after study was not possible at the time of writing this paper. Therefore, the total number of WWD acts and whether the wrong-way vehicles turned around were used to evaluate the performance of these devices. A WWD act occurs when a vehicle drives the wrong way and a turnaround occurs when the wrong-way driver self-corrects himself/herself by turning around after seeing the RFBs at the treated ramps. The WWD acts do not include vehicles reversing; bicyclists driving the wrong way; or mowing crews, utility vehicles, or emergency vehicles driving the wrong way on instrumented ramps. The total number of WWD acts that turned around due to the RFBs was verified by pictures taken from the cameras located on the "Wrong Way" signs and/or closed-circuit television cameras monitored by the Regional Traffic Management Center.

Figure 3. Example of a wrong-way driver self-correcting due to RFBs.

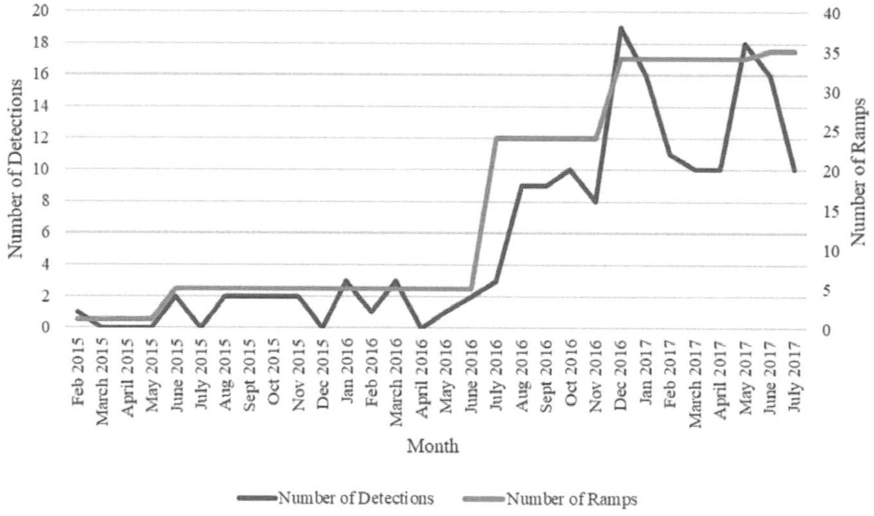

Figure 4. Total Monthly WWD Acts (February 21, 2015 – July 24, 2017).

Images for Evaluating RFBs' Performance

To confirm a detected WWD act, UCF researchers had to check all the images captured by the RFB cameras. In addition to the two cameras attached to the RFBs, three of the 35 treated ramps were equipped with a third overview camera. If a wrong-way vehicle continues and attempts to enter the mainline or turns around after the second camera location, the vehicle is captured by the overview camera. Each of these three cameras generates 7–20 pictures when the RFB detects any wrong-way vehicle on the exit ramp. These pictures can be viewed from a website created by the camera manufacturer. Figure 3 shows example images for a single wrong-way vehicle captured by the RFB cameras showing the vehicle turning around.

WWD Acts and Turnarounds Due to RFBs

Since the first RFBs were implemented in February 2015, there have been 170 WWD acts detected by the CFX RFBs through July 24, 2017. Figure 4 shows the monthly trends of the WWD acts for all CFX RFB sites. The total number of WWD acts in each month was very low through July 2016, as only five ramps had RFBs. Starting in August 2016, the number of monthly WWD acts increased significantly due to the higher number of equipped sites.

Table 2. WWD Detections and Turnarounds at all 35 CFX Sites and 17 FTE Sites

WWD countermeasure	Period covered	Time of day	Total WWD detections	Turnarounds	Turnaround percentage	Ratio of nighttime-to-daytime turnaround percentages
RFBs	Feb 2015–July 24, 2017	Daytime	54	41	75.9	1.02
		Nighttime	116	90	77.6	
		Total	170	131	77.1	
LEDs	Oct 2014–June 30, 2017	Daytime	18	4	22.2	0.45
		Nighttime	40	4	10	
		Total	58	8	13.8	

Note: RFBs = Rectangular Flashing Beacons; LEDs = Light Emitting Diodes.

Table 2 summarizes WWD detections and turnarounds for all 35 CFX sites through July 24, 2017. It also shows the number of turnarounds for the 17 FTE sites equipped with LEDs in South Florida. Since the RFB sites have two cameras to confirm if wrong-way vehicles turn around and the LED sites have only one camera, the ability to confirm turnarounds with LEDs is lower. The actual number of LED turnarounds could be higher than the confirmed number. This table shows that 131 of the total 170 WWD acts (77.1%) at the RFB-equipped ramps were turnarounds compared with 13.8% at the LED sites. This higher percentage of turnarounds for RFBs was one of the factors that led FTE to install double sets of RFBs on 18 exit ramps on their portion of SR 417 in Central Florida in June 2017. Table 2 indicates that the percentage of turnarounds for both daytime and nighttime WWD acts was higher at the RFB sites (75.9% for daytime and 77.6% for nighttime) than at the LED sites (22.2% for daytime and 10% for nighttime). These results also indicate that the RFBs were better than the LEDs at prompting nighttime drivers to turn around, as the ratio of nighttime-to-daytime turnarounds was more than twice as high (1.02 for RFB sites and 0.45 for LED sites).

Survey on WWD Countermeasures

Survey Methodology Applied to Both RFBs and LEDs

Three online surveys were created to gauge the opinions of transportation professionals, law enforcement officers, and road users about the effectiveness of WWD countermeasures. The first survey obtained responses from 247 FHP officers and 5 FDOT officials. The second (FTE customers and Florida interstate users) and third surveys (CFX customers) each had 900 random participants. Respondents of these surveys were asked to watch two short videos of WWD countermeasures. The first video showed one set of LED "Wrong Way" signs as installed by the FTE (example shown in Figure 1*a*) and the second video showed two sets of RFB "Wrong Way" signs as installed by CFX (example shown in Figure 1*b*). Both videos were taken from the perspective of a driver going the wrong way up an exit ramp and triggering the devices, took place in a similar nighttime setting, and had the vehicles traveling at roughly the same speed. Once the respondents finished watching these two videos, they were asked to choose their preferred option to alert and prevent wrong-way drivers. They were also asked to provide the reasons for choosing their preferred WWD countermeasure.

Survey Results

The first survey was designed to obtain agency opinions on WWD countermeasures. The FHP officers were asked about their thoughts on how to best avert WWD on limited-access roadways. This open-ended response question contained a variety of responses, but popular ones included better signs with flashing lights or sirens, tire deflation devices, more signs, bigger or better signs, alert and detection systems (cameras and sensors), and increased law enforcement. When shown the videos of the LED and RFB signs, 70% of the surveyed FHP officers preferred the RFBs over the LEDs. The officers who preferred the RFBs mainly did so due to both the additional set of signs and better flashing (70%), whereas 16% preferred the additional set of signs only and 7% preferred the flashing only. For the respondents who chose the LED signs, 53% said that both the external lighting and flashing were better, 19% said the external lighting was better, and 11% said the flashing was better. Like FHP officers, a majority (four out of five) of FDOT respondents preferred the RFB signs, with three preferring them due to the extra set of signs and one preferring them due to both the extra set of signs and better flashing.

Although the first survey focused on an agency perspective of these countermeasures, the second and third surveys focused on a road user perspective. A total of 72% of the FTE and Florida Interstate user respondents and 76% of the CFX respondents preferred RFBs over the LEDs. For the CFX respondents who preferred RFBs, 67% preferred them due to both the additional set of signs and better flashing, 18% preferred them due to better flashing only, and 15% preferred them due to the additional set of signs only. These outcomes show that an extra set of LED signs, improvements to the external lighting or flashing pattern at RFB sites, or the use of RFB-equipped signs instead of or in addition to the LED signs could be more efficient at reducing WWD.

Summary and Conclusion

"Wrong Way" signs with LEDs and RFBs are two types of advanced technology systems deployed to detect and prevent WWD at freeway exit ramps. The main objectives of this research were to analyze data collected from locations where these two countermeasures were implemented in Florida and determine potential improvements to these systems. Although there is other past and current research focused on examining LED-enhanced signs, this is the first research to evaluate the effectiveness of the double sets of RFBs as a WWD countermeasure. Evaluation of WWD countermeasure effectiveness is important to help agencies select the most efficient treatment for combating WWD on limited-access facilities and make sound investments by deploying the most effective treatment.

Based on the analysis of more than two and a half years' worth of data and observations collected in Florida, it was found that both the LEDs and RFBs have resulted in a reduction in WWD activities in the test corridors as of June 2017. A significant reduction in WWD events at the FTE treated sites in South Florida was found due to the implementation of LED signs. For instance, a 49% reduction in WWD 911 calls (at $\alpha = .01$) and a 38% reduction in combined WWD 911 calls and citations (at $\alpha = .05$) were found from this study. A similar study had been done by TTI for LED signs installed in Texas, which found a 38% reduction in the WWD 911 calls (*12*).

At CFX treated sites in Central Florida, 77.1% of the total WWD acts had wrong-way drivers self-correct because of RFBs compared with 13.8% that self-corrected at the FTE LED sites. One potential reason for this large difference is that the RFBs had an extra camera (oncoming) compared with the LEDs. This additional (oncoming) camera on the RFBs gives an additional field of view of the wrong-way vehicles that turn around on the exit ramps. The percentage of turnarounds for the nighttime WWD acts (77.6%) was slightly higher than the percentage of turnarounds for the daytime WWD acts (75.9%) for RFBs, whereas the percentage of turnarounds for the daytime WWD acts (22.2%) was significantly higher than the percentage of turnarounds for the nighttime WWD acts (10%) for LEDs. The nighttime-to-daytime turnaround ratio of 1.02 for the RFBs compared with 0.45 for the LEDs suggests that RFBs work more effectively in nighttime conditions.

The three surveys conducted regarding these two WWD countermeasures all had similar results, with 70% or more of respondents preferring RFBs over LED signs. RFBs were preferred by 70% of law enforcement officers, 80% of FDOT district officials, 72% of FTE customers and interstate users, and 76% of CFX customers. Most of these participants chose the RFBs over the LEDs because of the extra set of signs and better flashing.

Although these countermeasures reduce WWD on the highway, wrong-way drivers that are suicidal or intentional might not respond to these countermeasures and continue onto the highway mainline. For these types of drivers, another solution is needed. However, by integrating LED and RFB countermeasures with the TMC, agencies can notify right-way drivers of detected WWD events and reduce the chance of a WWD crash. Both FTE and CFX are working to incorporate the WWD alerts from these countermeasures into the regional SunGuide TMC software reporting system.

This paper confirms the effectiveness of these two advanced technology WWD countermeasures. The reduction of WWD events and high percentage of WWD turnarounds at the ramps where the LED and RFB signs were installed indicate a reduction in the risk of WWD crashes at these locations. Based on the survey results and RFB evaluation, improvements could be made to the LEDs, including an additional set of signs and multiple cameras, which could improve the turnaround percentage of these devices. The performance of the RFBs could potentially be improved by providing better external lighting and modifying the flashing pattern. WWD event data, including WWD crashes, citations, and 911 calls, as well as detection data from these two countermeasures, will continue to be collected to monitor and evaluate the effectiveness of these devices.

Acknowledgments

The research and results discussed in this paper are from projects funded by the Florida Turnpike Enterprise (FTE) and Central Florida Expressway Authority (CFX).

Author Contributions

The authors confirm contribution to the paper as follows: study conception and design: M.I.K., H.A-D., A.S., J.H.R., and G.C.; data collection M.I.K., H.A-D., A.S., J.H.R., and G.C.; analysis and interpretation of results: M.I.K., H.A-D., A.S., J.H.R., and G.C.; draft manuscript preparation: M.I.K., H.A-D., A.S., J.H.R., and G.C. All authors reviewed the results and approved the final version of the manuscript.

References

1. CBS Pittsburgh. *Pa. Leads Nation in Wrong Way Traffic Fatalities*. CBS Broadcasting Inc., Pittsburgh, PA. February 8, 2013. http://pittsburgh.cbslocal.com/2013/02/08/pa-leads-nation-in-wrong-way-traffic-fatalities/. Accessed July 28, 2017.
2. Kittelson & Associates, Inc. *Statewide Wrong Way Crash Study*. Final Report. Prepared for Florida Department of Transportation (FDOT). FDOT Central Office. Project No. 12274.03 Financial Project ID: 190258-1-32-37, Tallahassee, FL. January 2015.
3. Rogers, J. H., Jr., , H. Al-Deek, A. Alomari, F. A. Consoli, and A. Sandt. Wrong-Way Driving on Florida Toll Roads: An Investigation into Multiple Incident Parameters and Targeted Countermeasures for Reductions. *International Journal of Engineering Management and Economics*, Vol. 5, No. 3–4, 2015, pp. 144–168.
4. Zhou, H., J. Zhao, M. Reisi Gahrooei, and P. A. Tobias. Identification of Contributing Factors for Wrong-Way Crashes on Freeways in Illinois. *Journal of Transportation Safety & Security*, Vol. 8, No. 2, 2016, pp. 97–112.
5. Zhou, H., J. Zhao, M. Pour-Rouholamin, and P. A. Tobias. Statistical Characteristics of Wrong-Way Driving Crashes on

Illinois Freeways. *Traffic Injury Prevention*, Vol. 16, No. 8, 2015, pp. 760–767.
6. Pour-Rouholamin, M., and H. Zhou. Analysis of Driver Injury Severity in Wrong-Way Driving Crashes on Controlled-Access Highways. *Accident Analysis & Prevention*, Vol. 94, 2016, pp. 80–88.
7. National Transportation Safety Board. *Wrong-Way Driving*. Highway Special Investigation Report NTSB/SIR-12/01. National Transportation Safety Board, Washington D.C., 2012.
8. Gross, F., B. Persaud, and C. Lyon. *A Guide to Developing Quality Crash Modification Factors*. Report No.FHWA-SA-10-032. U.S. Department of Transportation, Washington, D.C., 2010.
9. Tamburri, T. *Wrong-Way Driving Accidents Are Reduced*. Highway Research Record #292. California Division of Highways, 1969.
10. Leduc, J. L. K., and S. L. Attorney. *Wrong-Way Driving Countermeasures*. Connecticut General Assembly, Hartford, Connecticut. September 22, 2008. https://www.cga.ct.gov/2008/rpt/2008-r-0491.htm. Accessed July 28, 2017.
11. Campbell, B. E., and P. B. Middlebrooks. *Wrong-Way Movements on Partial Cloverleaf Ramps*. Final Report. No. FHWA-GA-88-8203. Georgia Department of Transportation, Atlanta, 1988.
12. Finley, M. D., S. P. Venglar, V. Iragavarapu, J. D. Miles, E. S. Park, S. A. Cooner, and S. E. Ranft. *Assessment of the Effectiveness of Wrong Way Driving Countermeasures and Mitigation Methods*. Report No. FHWA/TX-15/06769-1. Texas A&M Transportation Institute, 2014.
13. Cooner, S. A., A. S. Cothron, and S. E. Ranft. *Countermeasures for Wrong-Way Movement on Freeways: Overview of Project Activities and Findings*. Publication FHWA/TX-04/4128-1. Texas Transportation Institute, College Station, 2004.
14. Athey Creek Consultants. *Countermeasures for Wrong-Way Driving on Freeways*. Enterprise Transportation Pooled Fund Study TPF-5 (231), West Linn, OR. September 2016.
15. Shepard, F. D. Evaluation of Raised Pavement Markers for Reducing Incidences of Wrong-Way Driving. *Transportation Research Record: Journal of the Transportation Research Board,* 1976. 597: 41..
16. Finley, M. D., J. D. Miles, and E. S. Park. Closed-Course Study to Examine the Effect of Alcohol-Impairment on a Driver's Ability to Identify and Read Signs. *Transportation Research Record: Journal of the Transportation Research Board,* 2017. 2660: 86-93.
17. Finley, M. D., K. N. Balke, R. Rajbhandari, S. T. Chrysler, C. S. Dobrovolny, N. D. Trout, P. Avery, D. Vickers, and C. Mott. *Conceptual Design of a Connected Vehicle Wrong-Way Driving Detection and Management System*. Report No. FHWA/TX-16/0-6867-1 2016. Texas A&M Transportation Institute, College Station, April 2016.
18. Johnson, R., and C. Harvey. *Harris County Toll Road Authority Incident Management's Rapid Response & Rapid Removal*. ITS Texas Annual Meeting, Houston, 2013. http://itstexas.org/sites/itstexas.org/files/presentations/2013%202B%20HCTRA%20Incident%20Management%E2%80%99s%20Rapid%20Response%20&%20Rapid%20Removal.pdf. Accessed July 28, 2017.
19. Venglar, S. P., and B. G. Fariello. Efforts to Reduce Wrong-Way Driving: A Case Study in San Antonio, Texas. Presented at 93rd Annual Meeting of the Transportation Research Board, Washington, D.C., 2014.
20. Florida Department of Transportation. *FDOT District Seven Wrong Way Driving (WWD) Countermeasures 2017 Update and Lessons Learned*. Florida Department of Transportation, 2017.
21Ozkul, S., and P.-S. Lin. Evaluation of Red RRFB Implementation at Freeway Off-Ramps and Its Effectiveness on Alleviating Wrong-Way Driving. *Transportation Research Procedia*, Vol. 22, 2017, pp. 570–579.
22. Hunter, W. W., R. Srinivasan, and C. A. Martell. *Evaluation of the Rectangular Rapid Flashing Beacon at A Pinellas Trail Crossing in St. Petersburg, Florida*. Final Report Submitted to FDOT. University of North Carolina, Highway Research Safety Center, October 2009.
23. CBS SF Bay Area. *New Traffic Radar Aims To Stop Wrong-Way Drivers Before They Start*. CBS Broadcasting Inc., San Francisco, CA. June 02, 2017. http://sanfrancisco.cbslocal.com/2017/06/02/new-traffic-radar-aims-to-stop-wrong-way-drivers-before-they-start/. Accessed July 28, 2017.
24. Easterling, J., and E. Gordin. *FTE Wrong-Way Driving Pilot Project Shows Promise*. Florida Department of Transportation's Traffic Engineering and Operations Newsletter., 2015, p. 1.
25. Sandt, A., H. Al-Deek, and J. H. Rogers Jr. Identifying Wrong-Way Driving Hotspots by Modeling Crash Risk and Assessing Durations of Wrong-Way Driving Events. *Transportation Research Record: Journal of the Transportation Research Board*, 2017. 2616: 58–68.
26. Rogers, J. H., Jr., , H. Al-Deek, A. H. Alomari, E. Gordin, and G. Carrick. Modeling the Risk of Wrong-Way Driving on Freeways and Toll Roads. *Transportation Research Record: Journal of the Transportation Research Board*, 2016. 2554: 166–176.
27. Al-Deek, H., J. Rogers, A. Sandt, and A. Alomari. *Evaluating the Wrong-Way Driving Incidents Problem on the Florida's Turnpike Roadway System, Phase-1 Study*. Final Report Submitted to FTE, Ocoee, FL. March 2016.
28. Rogers, J. *Evaluating Wrong-Way Driving for Florida Interstates and Toll Road Facilities: A Risk-Based Investigation, and Countermeasure Development*. PhD dissertation. Department of Civil, Environmental, and Construction Engineering, University of Central Florida, 2016.
29. Rogers, J. H., A. Sandt, H. Al-Deek, A. H. Alomari, N. Uddin, E. Gordin, C. Dos Santos, J. Renfrow, and G. Carrick. Wrong-Way Driving: Multifactor Risk-Based Model for Florida Interstates and Toll Facilities. *Transportation Research Record: Journal of the Transportation Research Board*, 2015. 2484: 119–128.
30. Florida Department of Transportation. *Interchange Report*. Florida Department of Transportation. http://www.fdot.gov/planning/statistics/hwydata/interchange.pdf. Accessed July 28, 2017.
31. *Licensed Drivers by Age, Sex and County in Florida*. The Florida Department of Highway Safety and Motor Vehicles, 2017. https://www.flhsmv.gov/resources/driver-and-vehicle-reports/. Accessed July 28, 2017.

The Standing Committee on Freeway Operations (AHB20) peer-reviewed this paper (18-04712) and selected it to receive its 2018 Best Paper Award.

TRR
JOURNAL OF THE TRANSPORTATION RESEARCH BOARD

Article

Transportation Research Record
2018, Vol. 2672(14) 106–116

DOI: 10.1177/0361198118778939
journals.sagepub.com/home/trr
SAGE

Wrong-Way Driving: A Regional Transportation Systems Management and Operations Approach to a Regional Problem

Md Omar Faruk[1], Haitham Al-Deek[1], Adrian Sandt[1], and John H. Rogers, Jr[1]

Abstract

Wrong-way driving (WWD) has been problematic on United States highways for decades despite its rare occurrence. Since WWD crashes are rare, recent researchers have studied WWD non-crash events such as WWD 911 calls and WWD citations to understand the overall nature and trend of WWD. This paper demonstrates the regional nature of the WWD problem and proposes regional transportation systems management and operations (TSM&O) solutions to combat this problem. Specifically, it was found that 11% of all WWD multi-data events (e.g., multiple 911 calls for the same WWD event) involved travel from one county to another. Additionally, 30% of all WWD single-data and multi-data events occurred at or near interchanges between two limited access highways in counties with multiple operating agencies. This indicates that a significant proportion of WWD events could potentially travel from one limited access facility to another. Moreover, 28% of WWD events occurred on limited access facilities shared by multiple operating agencies. To emphasize the regional nature of WWD, this paper determined the vulnerable demographic groups in different regions of Florida by developing WWD crash and citation prediction models. The models' findings indicate that certain demographic groups (e.g., elderly drivers) increase WWD risk. The models' results can be used to improve driver education and increase law enforcement presence in high risk WWD locations. Regional TSM&O solutions, such as coordination and communication among agencies and traffic management centers (TMCs), law enforcement co-location with TMCs, and strengthening statewide TSM&O programs to manage WWD events are also proposed.

Limited access roadways (such as interstates and toll roads) are designed to provide high speed mobility and safe traveling by eliminating potential crashes. Despite modern design standards and introduction of intelligent transportation systems technologies, limited access roadways still experience safety challenges; one such challenge is wrong-way driving (WWD). WWD occurs when a driver intentionally or unintentionally (due to confusion, intoxication, or other reasons) drives on the wrong side of divided highways against the direction of mainline traffic flow. WWD crashes account for only 3% of all crashes (*1*). However, WWD crashes draw more media attention due to high fatality rates and can worry road users (*2*). A recent United States study found that, from 2004 to 2013, there were approximately 1.34 fatalities per fatal WWD crash compared to 1.10 fatalities in all other fatal crashes (*3*).

Since WWD crashes are rare, it can be difficult to understand WWD by analyzing WWD crashes alone. Previous studies conducted by the University of Central Florida (UCF) on limited access highways in Florida showed that WWD non-crash events, such as WWD citations and WWD 911 calls, can be used as surrogate measures for WWD crashes due to their more frequent occurrence than WWD crashes (*2, 4–6*). As crashes are rare in nature, it is not always recommended to wait until crashes happen before introducing countermeasures. Study findings by UCF using non-crash WWD data have been adopted by transportation agencies to implement WWD countermeasures in South Florida and Central Florida (*2, 4–6*).

[1]Department of Civil, Environmental, and Construction Engineering, University of Central Florida (UCF), Orlando, FL

Corresponding Author:
Address correspondence to Haitham Al-Deek: Haitham.Al-Deek@ucf.edu

This paper investigates the regional aspects of the WWD problem, such as how many counties a wrong-way (WW) driver travels through from when they are first reported to when they are last reported, or stopped by law enforcement, or crash. A WW driver can also travel from one limited access roadway to another limited access roadway in the same county. This can be a regional problem if the two roadways are operated by different agencies. In some counties, multiple agencies operate the same limited access highways, so a WWD event could cross operational boundaries easily and quickly. For these reasons, WWD needs to be dealt with in a regional manner. This paper also provides some regional transportation systems management and operations (TSM&O) solutions to deal with this regional problem.

To enhance the regional approach, this paper also examines the demographic characteristics of WW drivers. It introduces a regional analysis concept to predict WWD crashes in different regions in Florida using both WWD non-crash events and driver demographic information. Florida has a wide variety of ethnicity and age groups living in various parts of the state, so this type of analysis can help traffic agencies throughout the state determine the potential WWD risk related to various demographic groups. These demographic differences need to be considered for effective WWD countermeasure implementation. This regional-based approach investigating the effects driver demographics and non-WWD crash events have on WWD crashes and WWD citations has never been attempted before.

Literature Review

WWD research was started by the California Department of Transportation (Caltrans) in the 1950s with the study of median-related crashes on freeways (*7*). California first started using cameras and road tubes in 1967 to detect WWD incidents and identify locations with higher WWD crash density (*8*).

Studies conducted in Texas and Illinois found that urban areas experience more WWD crashes than rural areas (*9–11*). The Texas study found that drivers younger than 35 years and over 65 years were overrepresented in crashes, as were Hispanic drivers. Additionally, most WWD crashes occurred during nighttime (*9*, *10*). The Illinois study found that approximately 60% of WWD crashes were caused by drivers driving under influence (*11*). A North Carolina study also found that intoxicated drivers and elderly drivers were overrepresented in WWD crashes (*12*). Younger drivers typically caused crashes due to distraction (*13*), whereas older drivers caused crashes because of physical illness, such as dementia, or confusion (*2*).

A study of WWD crashes on French divided highways from 2008 to 2012 identified the factors that differentiate WWD crashes and non-WWD crashes on the same road over the same period (*13*). It was found that WWD crashes are more likely to occur during nighttime and WW drivers are more likely to be older, intoxicated, local, and drive older vehicles. Lathorp et al. reviewed fatalities caused by WWD crashes on New Mexico's interstates occurring from 1990 to 2004 and compared them with other fatal non-pedestrian crashes (*14*). The analysis found that fatal WWD crashes are more likely to occur during darkness and involve intoxicated drivers. In San Antonio, Texas, there were 185 reports of WWD crashes and 358 WWD 911 calls in 2011 (*15*). 80% of the WWD events happened at night, with 45% occurring between 2 a.m. and 4 a.m. Another Texas study conducted a motorist survey to assess motorists' understanding of WWD warning messages (*16*). The survey findings supported the use of two WWD warning messages and how to display them on Dynamic Message Signs (DMSs) to alert right-way drivers.

UCF started its WWD research by surveying drivers on their WWD experiences and examining WWD crashes, citations, and 911 dispatch calls in the Central Florida area (*6*, *17*). Applying a new methodology of combining WWD crash and non-crash events, Rogers at al. ranked the counties and limited access routes in counties with high WWD risk value (*4*). The study found that counties with larger metropolitan areas in Florida (Miami-Dade, Orange, Broward and Palm Beach) were ranked highest in terms of WWD risk value (*4*). It also ranked the interstates and toll roads in these counties with high WWD risk (*4*). An extended study by the same researchers identified high-risk roadway segments (hotspots) at the microscopic level in South Florida (*2*). Sandt et al. extended the WWD hotspot modeling to examine limited access facilities in Central Florida (*5*). Besides using WWD citations, 911 calls, and geometric features of interchanges and traffic volume, the study also identified WWD hotspots based on WWD event durations obtained from the regional traffic management center (TMC) in Orlando, Florida (*5*). The identified hotspots in Central Florida obtained by both techniques were found to be similar (*5*).

Compared to research on WWD events, very little research has been done on regional TSM&O strategies to reduce WWD and provide WWD warning messages to drivers. Existing regional TSM&O strategies include a variety of management and operations solutions, such as incident management, work zone management, active traffic management, planned special event management, and road weather management (*18–20*). Most of the regional TSM&O research emphasized regional partnerships and building stronger links between planning and operations (*19*, *21*). Although regional TSM&O strategies are gaining significance by providing more economical alternatives, consideration of a regional TSM&O approach is still non-existent in some areas, such as WWD. This study, which is an extension of previous research by UCF, deals with the regional and demographic aspects of WWD and fills the gap in existing regional TSM&O research on WWD events. It shows how WWD is a regional problem and provides suggestions that agencies can adopt to reduce WWD and law enforcement response time on a regional level.

Research Objectives

The primary objectives of this research are to:

1) Study the regional aspects of WWD by examining the different counties, limited access facilities, and/or jurisdictions a WW driver travels through before he/she either crashes, is apprehended by law enforcement, or self-corrects his/her WWD behavior.
2) Analyze and model the demographic factors of drivers to find which driver factors influence the number of WWD crashes and citations in different regions in Florida.
3) Propose regional TSM&O solutions to the WWD problem. These proposed solutions are conceptual in nature and include better coordination and communication among local agencies to manage the WWD problem. Also, these solutions are compatible with existing regional TSM&O programs practiced nationwide in the United States. Furthermore, the findings from this paper can be used to guide local and state agencies in determining the counties or populations with higher risk of WWD events and taking necessary and appropriate steps to combat this WWD problem.

Research Methodology

To achieve the three mentioned research objectives, the following steps were needed:

1) Geo-locating WWD crashes, WWD citations, and WWD 911 calls on limited access roadways throughout Florida for the years 2011–2015 and removing the overlap among these events.
2) Determining the number of WWD multi-data events (events with either multiple WWD 911 calls, 911 calls and citations, or 911 calls and crashes) that traveled from one county to another or from one limited access facility to another.
3) Determining the number of WWD multi-data and single-data events (events with only one WWD 911 call, one citation, or one crash) occurring at the interchanges between two limited access facilities.
4) Determining the proportion of WWD events on limited access facilities shared by multiple agencies in counties where multiple agencies operate limited access facilities.
5) Exploring and collecting various demographic data to find which factors might increase the number of WWD events.
6) Developing statistical models to predict the number of WWD crashes and citations on limited access highways in each county of Florida using the driver demographic variables previously collected.
7) Studying and exploring regional TSM&O practices and research on incident management agreements among agencies (such as state departments of transportation, regional transportation agencies, local police departments, local emergency response departments etc.) available statewide and nationwide to determine the documents which are most pertinent to high profile incident management like WWD.
8) Reviewing existing standard operations procedures regarding WWD detection, alert, and incident response practiced by different agencies in Florida and finding areas for improvement.

Data Collection

Various data were needed for this research, including Florida county data, WWD event data, and driver demographic information.

1) Selection of studied counties.

Among the 67 counties in Florida, only 45 of them have limited access highways. Therefore, only these 45 counties were considered in this study. The county boundaries were obtained from the U.S. Census Bureau Tiger-Line data (*22*).

2) Mapping WWD events.

WWD events were mapped on Florida's limited access highways using provided longitude and latitude coordinates. The limited access highways include I-4, I-10, I-75, I-95, I-110, I-195, I-275, I-295, I-375, I-395, I-595, SR 91, SR 112, SR 408, SR 414, SR 417, SR 429, SR 528, SR 451, SR 570, SR 589, SR 618, SR 821, SR 826, SR 836, SR 869, SR 874, SR 878, and SR 924.

3) Driver demographic information.

Table 1. Summary of Multiple WWD Events Type and Number of Counties Traveled by WW Driver

Multiple WWD event type	Same county (first county)	Neighboring next county (second county)	Crossed one county (ended up in third county)
Multiple WWD calls events	435 (89.9%)	37 (7.6%)	12 (2.5%)
WWD calls and citations	39 (83%)	7 (14.9%)	1 (2.1%)
WWD calls and crashes	31 (93.9%)	2 (6.1%)	NA
Grand total	505 (89.5%)	46 (8.2%)	13 (2.3%)

Note: NA = not applicable.

Table 2. WWD Multi-data Events at or Near an Interchange between Two Limited Access Highways Operated by Multiple Agencies

County	Crossed on to another road	Stayed on the same road
Orange	2 out of 31 (6.5%)	10 out of 31 (32.3%)
Seminole	NA	1 out of 5 (20.0%)
Osceola	NA	2 out of 13 (15.4%)
Broward	3 out of 88 (3.4%)	15 out of 88 (17.0%)
Hillsborough	3 out of 56 (5.4%)	13 out of 56 (17.1%)
Miami-Dade	5 out of 53 (9.4%)	12 out of 53 (22.6%)
Polk	1 out of 20 (5.0%)	5 out of 20 (25.0%)
Total	14 out of 248 (5.7%)	58 out of 248 (23.4%)

Table 3. WWD Single-data Events at or Near an Interchange between Two Limited Access Highways Operated by Multiple Agencies

County	911 Calls	Citations	Crashes
Orange	9 out of 32 (28.1%)	16 out of 50 (32.0%)	6 out of 18 (33.3%)
Seminole	4 out of 7 (57.1%)	3 out of 8 (37.5%)	3 out of 8 (37.5%)
Osceola	2 out of 11 (18.2%)	3 out of 17 (17.7%)	1 out of 4 (25.0%)
Broward	21 out of 81 (25.9%)	25 out of 93 (26.9%)	5 out of 21 (23.8%)
Hillsborough	21 out of 59 (35.6%)	12 out of 34 (35.3%)	5 out of 19 (26.3%)
Miami-Dade	17 out of 60 (28.3%)	45 out of 118 (38.1%)	13 out of 50 (26.0%)
Polk	8 out of 21 (38.1%)	5 out of 22 (22.7%)	1 out of 6 (16.7%)
Total	82 out of 271 (30.3%)	109 out of 342 (31.9%)	34 out of 126 (27.0%)

Table 4. WWD Events on the Same Limited Access Highways Operated by Multiple Agencies

Roadways operated by multiple agencies	911 Calls	Citations	Crashes
SR 924 (MDX) & I-75 (FDOT) in Miami-Dade	15	5	6
SR 112 (MDX) & I-195 (FDOT) in Miami-Dade	8	3	3
SR 836 (MDX) & I-395 (FDOT) in Miami-Dade	11	28	8
SR 417 (FTE in Osceola and Seminole; CFX in Orange)	18	18	10
SR 429 (FTE in Osceola and Orange; CFX in Orange)	15	9	1
SR 528 (FTE and CFX in Orange; FDOT in Brevard)	23	14	10
Total	90 out of 285 (31.6%)	77 out of 429 (18.0%)	38 out of 140 (27.1%)

Note: MDX = Miami-Dade Expressway Authority; FDOT = Florida Department of Transportation; FTE = Florida Turnpike Enterprise; CFX = Central Florida Expressway Authority.

Demographic information for each county was used to capture the regional variation in WWD events among the counties. Licensed driver information for each county was used instead of population information to better represent the number of potential drivers on the road. Florida Highway Safety and Motor Vehicles (HSMV) data was collected from 2011 to 2015 to determine the number and percentages of licensed drivers in different age groups (*23*). Population information for major ethnic groups was collected from the U.S. Census

Bureau—American FactFinder data for 2011–2015 (*24*).

Regional WWD Analysis

The concept of a region-based analysis that considers each county as a unit stems from the notion that most WWD events end in the same county where they originate. To prove this point, all WWD events with multiple 911 calls, 911 calls and citations, or 911 calls and crashes were investigated to determine whether they ended in the same county where they started. This analysis was performed on all WWD multi-data events from 2011 to 2015. However, the origin of a WW driver is often unknown, as it is impossible to determine with 100% certainty where the driver first entered WW, so it was assumed that they started going WW at the closest interchange to the first data point of the event. First, the WWD events with multiple 911 calls but no citations or crashes were analyzed. Among these multi-call events, 89.9% ended in the same county where they originated, 7.6% traveled to a neighboring county, and 2.5% crossed one county and ended up two counties away from their origin. The first row in Table 1 shows the summary of all multi-call events and the number of counties crossed before last recorded.

The remaining rows of Table 1 summarize the county starting and ending points for the other types of WWD multi-data events (911 calls and citations, and 911 calls and crashes) and the total for all multi-data events. In total, 89.5% of these events had the first and last data points in the same county, while 8.2% ended in an adjacent county and 2.3% ended up two counties away. These statistics indicate that WWD events can be analyzed as a county-based regional phenomenon.

Only 7 out of 45 counties (Orange, Seminole, Osceola, Broward, Hillsborough, Miami-Dade, and Polk) have limited access highways operated by multiple agencies. For example, in Orange County, the Florida's Turnpike Enterprise (FTE) and the Central Florida Expressway Authority (CFX) operate the toll roads while interstates are operated by Florida Department of Transportation (FDOT). A WW driver can travel from one limited access facility under jurisdiction of one agency to another limited access facility under jurisdiction of another agency. Therefore, it is important to study the interchanges between two limited access facilities in the same county that are operated by multiple agencies. Table 2 shows the analysis of WWD multi-data events at or near these interchanges. In the above counties, 29.1% of WWD multi-data events occurred at or near these interchanges, and 5.7% of all WWD multi-data events traveled from one limited access facility to another. The single-data events near these interchanges were also analyzed; these results are shown in Table 3. These events could have had the WW vehicle traveling on both facilities, but this is unknown since there was only one WW data point for each of these events. About 30% of all WWD single-data events in the above counties occurred at or near interchanges between two limited access facilities, indicating that there is potential for WW vehicles to travel on multiple limited access facilities within the same county.

Even if WW vehicles stay on the same roadway, they could still cross operational boundaries, as multiple agencies can operate the same limited access facility in one or multiple counties. Table 4 shows WWD events on limited access facilities that are operated by multiple agencies. A WW driver could enter the facility where it is operated by one agency and keep driving WW on it through sections operated by another agency. For example, on 11/21/2014, a 911 call was recorded on the CFX section of SR 429 in Orange County at 23:39, then a second 911 call for the same vehicle was recorded on the FTE section of SR 429 in Osceola County at 23:50. Analysis of the WWD events on the indicated roadways shows that 31.6%, 18.0%, and 27.1% of all WWD 911 calls, citations, and crashes, respectively, in the indicated counties occurred on limited access facilities shared by multiple agencies. This shows the need to coordinate WWD events on a regional level between different neighboring agencies.

Data Preparation for Modeling

A statistical model was developed for predicting WWD crashes in each county using WWD citation, WWD 911 calls, and driver demographic information. Another statistical model was developed for predicting WWD citations in each county using WWD 911 calls and demographic information. For statistical modeling, it is necessary to remove the overlap among the multi-data point events. A market basket analysis was performed to carry out this task as was done in previous research (*2, 4, 5*). There were instances where 911 calls, citations, and crashes were reported for the same WW vehicle, but these were considered as one event to remove duplication before attempting statistical modeling. The data was organized so that every WW event was represented uniquely in developing statistical models.

After performing the market basket analysis, there were 249 crashes, 1671 citations, and 2496 WWD calls recorded from 2011 to 2015 on limited access roadways in Florida; these events are shown in Figure 1. Figure 1*a* shows the WWD crashes, Figure 1*b* shows the WWD citations, and Figure 1*c* shows the WWD 911 calls on limited access highway segments in each of the 45 selected counties. A maximum of 50 crashes was observed in

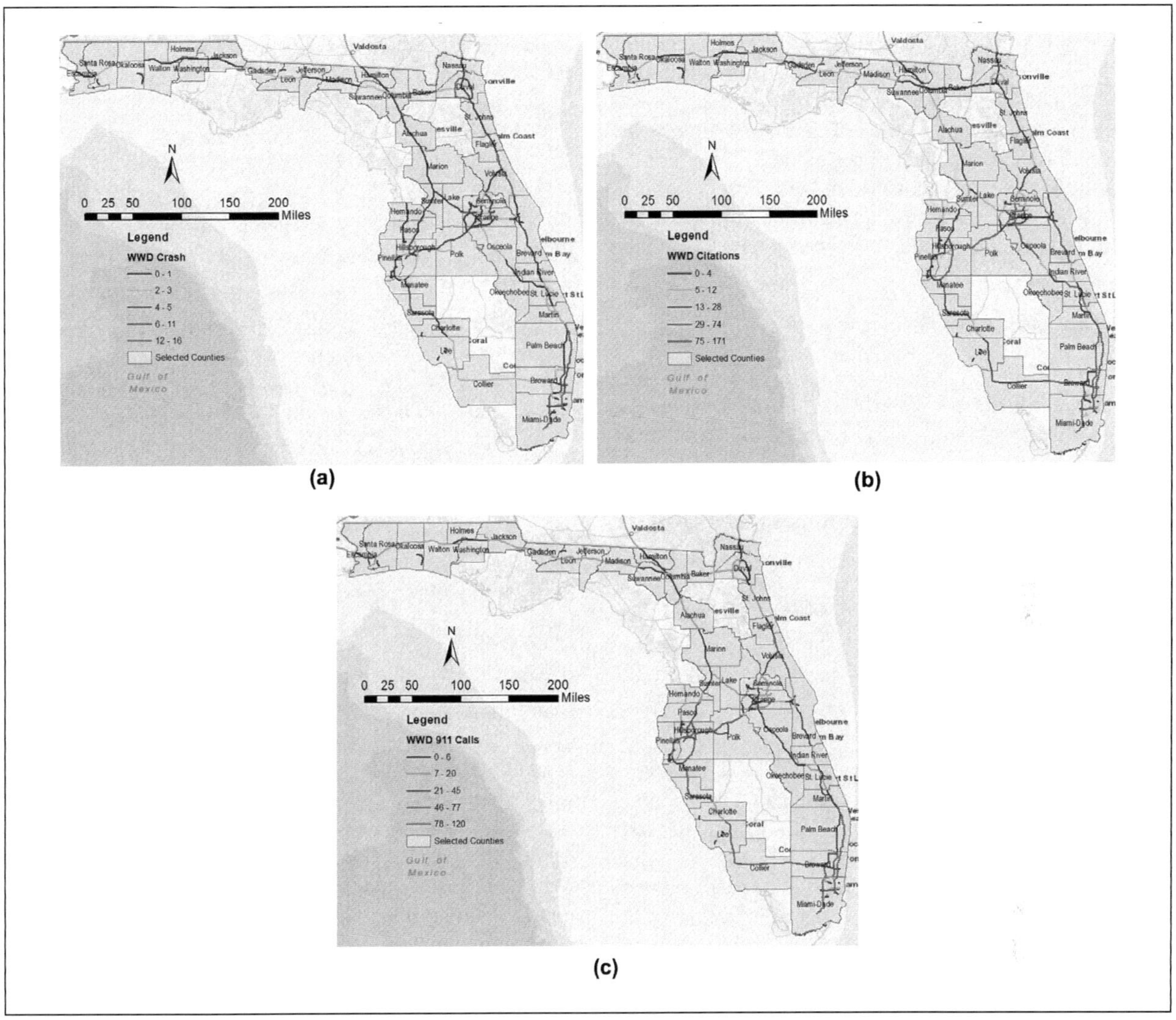

Figure 1. WWD events on roadway segments in each selected county, showing (*a*) crashes, (*b*) citations, and (*c*) 911 calls.
Note: 1 mile = 1.61 km.

Miami-Dade county and no crashes were observed in Alachua, Jackson, Jefferson, and St. Johns counties, which indicates a high variation in the number of crashes. The number of citations varied from 269 (Miami-Dade) to 1 (Washington) and the number of 911 calls varied from 360 (Hillsborough) to 1 (Holmes).

County type (urban/rural) data was another parameter investigated in this study. Among the 45 counties investigated, 33 (73.3%) were urban and 12 (26.7%) were rural. Another factor considered was daily vehicle miles traveled (DVMT) on limited access highways in the selected counties as a measure of exposure.

For the driver demographics, information on licensed drivers in each county was used. Driver age, gender, and ethnicity were the important factors that were considered. Data for these factors were collected for all years from 2011 to 2015 and a weighted average was taken for statistical modeling. The number of licensed drivers was divided into several groups by age: 15–30, 30–40, 40–50, 50–65, and above 65 years (*25*, *26*). To avoid over- or under-representation for different age and ethnic groups, the number of drivers in each age group and population of each ethnic group were expressed in percentages to provide standardized values.

Statistical Models

Statistical modeling was started by performing a univariate analysis on all potential candidate variables. Only significant variables from the univariate analysis were kept for further analysis and inclusion in statistical models. It was observed that citations and 911 calls were not normally distributed, so a logarithmic transformation was performed to make them normally distributed. Multicollinearity among the significant parameters was checked before keeping them in the model.

Since both the WWD crashes and citations have variances larger than their means, a negative binomial model is appropriate for modeling this type of data (*27*). The negative binomial model has a form of

$$\ln(\lambda) = \beta x + \varepsilon \quad (1)$$

where
ln = natural logarithm (*e* based logarithm);
λ = expected number of WWD events (crashes or citations) in a county;
β = vector representing parameters to be estimated;
x = vector representing the explanatory variables in each county; and
ε = error term in the model.
SAS 9.4 software package was used to produce the WWD crash and citation models (*28*).

Model 1: WWD Crash Model

The traditional negative binomial regression model is given as

$$ln(\mu) = \beta_0 + \beta_1 x_1 + \beta_2 x_2 + \ldots + \beta_p x_p \quad (2)$$

where the predictor variables $x_1, x_2, \ldots, x_p$ are given, and regression coefficients $\beta_0, \beta_1, \beta_2, \ldots, \beta_p$ are to be estimated.

Taking the exponential of both sides results in

$$\mu = \exp(\beta_0 + \beta_1 x_1 + \beta_2 x_2 + \ldots + \beta_p x_p) \quad (3)$$

Based on the determined significant variables, the WWD crash prediction model developed in this study will resemble the equation

$$\mu = \exp(\beta_0 + \beta_1 x_1 + \beta_2 x_2 + \beta_3 x_3 + \beta_4 x_4 + \beta_5 x_5 + \beta_6 x_6 + \beta_7 x_7) \quad (4)$$

where
exp = exponential;
μ = number of WWD crashes in a county over 5 years;
X_1 = log (number of citations in a county over 5 years);
X_2 = log (number of 911 calls in a county over 5 years);
X_3 = MDVMT (Million DVMT) in a county;
X_4 = county type (1 = rural, 0 = urban);
X_5 = % of licensed drivers in age group (15–30) in a county;
X_6 = % of licensed drivers in age group (50–65) in a county; and
X_7 = % of Hispanic population in a county.

The final crash prediction model with parameter estimates is given as

$$\text{WWD Crashes} = \text{Exp}(-7.5568 - 0.5951 \times \log(\text{citation}) + 1.1079 \times \log(911\ \text{calls}) + 0.0942 \times (\text{MDVMT}) - 0.5463 \times (\text{county type}) + 0.0835 \times (\%\ \text{of licensed drivers in age group 15 to 30 years}) + 0.2083 \times (\%\ \text{of licensed drivers in age group 50 to 65 years}) + 0.0186 \times (\%\ \text{of Hispanic population})) \quad (5)$$

Table 5 shows the model output, which indicates that WWD 911 calls, MDVMT, percentage of licensed drivers aged 50–65 years, and percentage of Hispanic population are significant at 5% significance level (*p*-value $<.05$), whereas WWD citations, county type, and percentage of licensed younger drivers aged 15–30 years are significant at 10% significance level (*p*-value $<.10$). Table 6 shows explanation of parameter statistics. Pearson chi-square and deviance chi-square tests show that the model is a good fit model with the selected parameters. The best model was selected based on the smallest AIC and AICc values. The model predicts that more WWD citations decrease the number of WWD crashes in a county whereas more 911 calls increase the number of crashes. This is consistent with findings of previous research conducted by UCF (*1, 4, 6*). Before modeling, it was expected that WWD crashes would increase with higher DVMT and percentages of younger and older drivers in a county. The model predictions are consistent with these prior expectations. Additionally, urban counties are more vulnerable to WWD crashes, which is also consistent with prior expectation. The final variable is the presence of Hispanic population, with higher percentages increasing the number of WWD crashes. This finding could be attributed to unfamiliarity with "Wrong Way" signs or language.

Model 2: WWD Citation Model

Another model was developed to predict the number of WWD citations in each county over the five-year period. WWD 911 call and demographic information were considered in developing this model. The model developed for predicting WWD citations has the functional form

$$\mu = \exp(\beta_0 + \beta_1 x_1 + \beta_2 x_2 + \beta_3 x_3 + \beta_4 x_4 + \beta_5 x_5) \quad (6)$$

where
$\beta_0, \beta_1, \beta_2, \beta_3, \beta_4, and\ \beta_5$ = regression coefficients;
exp = exponential;
μ = number of WWD citations in a county over 5 years;
X_1 = log (number of 911 calls in a county over 5 years);
X_2 = log (average DVMT);

Table 5. Florida Statewide Statistical Model for WWD Crash Prediction

Parameter	Estimate	Standard error	Wald chi-square	Prob > chi-square
Intercept	−7.5568	2.5688	8.65	0.0033
Log(citations)	−0.5951	0.3718	2.56	0.095*
Log (911 Calls)	1.1079	0.3901	8.07	0.0045**
MDVMT (millions DVMT)	0.0942	0.0328	8.24	0.0041**
Rural/urban (1 = rural)	−0.5463	0.3653	2.24	0.097*
Rural/urban (0 = urban)	NA	NA	NA	NA
% of licensed driver (Age 15–30 years)	0.0835	0.045	3.45	0.0634*
% of licensed driver (Age 50–65 years)	0.2083	0.0809	1.01	0.01**
% of Hispanic population	0.0186	0.0081	5.26	0.0218**

Note: NA = not applicable.
*Significant at 10% significance level (p-value <.10);
**Significant at 5% significance level (p-value <.05);
Goodness of fit statistics: Pearson chi-square = 33.3980 on 37 d.f. (degrees of freedom), p-value >.05;
Deviance chi-square = 36.3523 on 37 d.f., p-value >.05;
AIC = 195.8996;
AICc = 201.0424.

Table 6. Explanation of Parameter Statistics in Table 3*a*

Parameter statistic	Explanation
Wald chi-square	The test statistic from the Wald chi-square test is used to test the hypothesis that the corresponding parameter = 0 in the model vs parameter ≠ 0
Pearson and deviance chi-square	Goodness of fit statistics. If the p-value of these tests is greater than 0.05, the model is a good fitting model
AIC	Akaike Information Criterion is used for the comparison of models on the same sample. Smaller is better
AICc	AIC with a correction for finite sample size. Smaller is better

X_3 = % of licensed drivers in age group (50 -65) in a county; and
X_4 = % of Hispanic population in a county.

The final equation for predicting WWD citations is

$$\begin{aligned}&\text{WWD Citations} =\\&\text{Exp}(-9.6786 + 0.9979 \times \log(911\text{calls})\\&+ 1.2872 \times \log(\text{DVMT})\\&+ 0.0937 \times (\%\text{ of licensed drivers in age group 50 to 65 years})\\&+ 0.0227 \times (\%\text{ of Hispanic population}))\end{aligned} \quad (7)$$

The model output is shown in Table 7, which indicates that WWD 911 calls, DVMT, and percentage of Hispanic population are significant at 5% significance level and the percentage of licensed drivers aged 50–65 is significant at 10% significance level. The Pearson and Deviance chi-square tests show the model is a good fit model. The AIC and AICc values were used to pick the best model for predicting WWD citations in a county. The model predictions are consistent with the prior expectations that more 911 calls, higher DVMT, and increased presence of elderly and Hispanic population drivers positively influence WWD citations.

Regional TSM&O Possible Solutions for WWD

The analyses discussed in this research have proven that a WW driver can travel from the roadway facility of one agency to the roadway facility of another agency. Agencies that operate different portions of the same or adjacent roadways need to be equally prepared to alert or stop a WW driver and alert other nearby right-way drivers in the network. There can be a regional TSM&O approach where a TMC can monitor any installed WWD prevention and alert systems by coordinating all local and regional agencies who oversee the limited access highways in a region. For example, the TMC can have access to all DMSs on the limited access facilities in a region and can then alert right-way drivers when a WWD event is detected using the DMS of any or all agencies in the region, irrespective of the roadway where the WWD event takes place. Thus, limited access

Table 7. Florida Statewide Statistical Model for WWD Citation Prediction

Parameter	Estimate	Standard error	Wald chi-square	Prob > chi-square
Intercept	−9.6786	3.1094	9.69	0.0019
Log (911 calls)	0.9979	0.3474	8.25	0.0041**
Log(DVMT)	1.2872	0.4126	8.32	0.0039**
% of licensed driver (age 50–65 years)	0.0937	0.055	2.9	0.0885*
% of Hispanic population	0.0227	0.01	5.12	0.0236**

Note: *Significant at 10% significance level (p-value <.10);
**Significant at 5% significance level (p-value <.05);
Goodness of fit statistics: Pearson chi-square = 43.8113 on 40 d.f., p-value >.05;
Deviance chi-square = 41.1183 on 40 d.f., p-value >.05;
AIC = 336.5643;
AICc = 338.7749.

roadways of neighboring agencies can be coordinated by a single regional center.

Co-location of law enforcement agencies with TMCs, setting up specific communication protocols between the TMCs and law enforcement, or having two-way data connections between TMC computers and police computer-aided dispatch are possible regional TSM&O solutions that could be useful for combating WWD events. These measures would improve the response time by law enforcement, which can reduce WWD crashes by allowing law enforcement to stop a WW driver before a crash occurs. These TSM&O solutions could need to be evaluated using a benefit–cost analysis to determine which approach is the most beneficial for different regions. The results from this research could be used to determine which TMCs should get priority for installing these countermeasures. TMCs located in areas with higher frequencies of WWD events (crashes, citations, and 911 calls) or where vulnerable demographic groups are more present would be the preferred locations for these solutions. Examples of these above two solutions already exist in Orlando, Florida, and other parts of the country. The regional TMC located in Orlando is managed by FDOT District-5; it oversees all DMS and WWD activities for roads operated by CFX and FDOT District-5. Florida Highway Patrol (FHP) Orlando headquarters are also co-located within the same regional TMC facility.

For areas with multiple TMCs and multiple agencies, coordination among TMCs is crucial. If a WW driver originates in one TMC's jurisdiction and shortly travels to a roadway under another TMC's jurisdiction, the TMCs need to collaborate and communicate to notify law enforcement to stop the WW driver. For example, I-95 runs through the jurisdictions of FDOT District-2, District-5, District-4, and District-6. A WW driver could start on District-2's portion of I-95 and shortly end up on District-5's portion of I-95. If the TMCs of both districts coordinate and work together in stopping the WW driver, they can do it more efficiently than each TMC working alone.

FDOT's TSM&O program, which aims at increasing the delivery rate of a fatality-free and congestion-free transportation system, can contribute by laying the grounds for programs that incorporate many system-based and institutional strategies related to WWD. This program can put forward rules and regulations that all statewide agencies must follow to increase social awareness about WWD, implementing countermeasures to stop a WW driver and alerting right-way drivers. The collection of all TMC functions under one umbrella, like FDOT's TSM&O program, will allow for more holistic, system-wide management of regional WWD events. Similar programs could be beneficial in other states as well.

Summary and Conclusions

Most of the previous WWD studies only focused on WWD crashes, which limited the practical use of the research findings due to the rare occurrence of crashes. Using WWD non-crash events is a newer concept which has been proven to be successful in practical use by transportation agencies due to the wider availability of data. However, no research has explored the regional aspects of WWD events to find how many counties WW drivers might travel through, whether they stay on the same roadway or travel to another roadway under jurisdiction of another agency, and how to address these issues. This study explores the regional aspects of WWD and proposes potential TSM&O solutions to address these issues.

Multi-data WWD events were analyzed to find out how often WW drivers crossed regional, roadway, or operational agency boundaries. These analyses showed that 10.5% of WWD multi-data events traveled from one county to another county. Therefore, some WWD events could require communication and coordination between neighboring counties to effectively stop the WW driver before he/she causes a crash. There are also some counties where multiple agencies operate limited access highways. WWD multi-data event analysis showed that about 6% of all WWD multi-data events in these counties traveled on multiple limited access facilities through an interchange between the two limited access facilities whereas 23% stayed on the same roadway. WWD single-data events showed that about 30% of 911 calls, 32% of citations, and 27% of crashes in these same counties occurred at or near the interchanges between two limited access facilities. These findings indicate that WW drivers could potentially travel on multiple facilities under the jurisdiction of different agencies. Moreover, there are some limited access facilities that are operated by multiple agencies. Analysis of WWD events on these facilities showed that about 32% of 911 calls, 18% of citations, and 27% of crashes in counties with roadways that have multiple operational agencies occurred on these facilities. This indicates that a significant proportion of WWD events could start on one agency's roadway and end up on another agency's roadway.

This paper also investigated regional driver demographics to find the demographic groups most susceptible to WWD. Two negative binomial models were developed to predict the number of WWD crashes and citations in each county. The crash prediction model found that counties with higher percentages of licensed drivers aged 15–30 and 50–65 years had more WWD crashes, as did urban counties and counties with a higher percentage of Hispanic population. The model also predicts that more WWD 911 calls in a county (which indicates more WWD activity) tend to increase the number of WWD crashes, while more WWD citations in a county tend to decrease the number of WWD crashes since the WW driver is stopped before causing a crash. The citation model found similar results, with counties that had more WWD 911 calls and higher percentages of licensed drivers aged 50–65 years and Hispanic population having more WWD citations. This citation model is useful for counties with very few WWD crashes to determine what groups are more susceptible to WWD. Improving WWD education for these identified groups, as well as increasing law enforcement presence in areas where these groups are more abundant are some potential solutions to decrease WWD events.

To solve this regional WWD problem, a regional TSM&O approach is needed. This paper proposed several regional TSM&O solutions to prevent and mitigate WWD. Coordination among local agencies and a designated TMC controlling DMSs on all limited access highways in a region irrespective of agencies would help to promptly detect a WW driver and alert all right-way drivers. Co-location of law enforcement with regional TMC would be another way to shorten the response time and stop a WW driver prior to crash. Coordination and communication among statewide TMCs, especially the ones which have the same limited access highway running through them would also help to effectively stop a WW driver. Moreover, strengthening statewide TSM&O programs where a state department of transportation would lead to coordinate all agencies, and involving road users by raising social awareness, would help to proactively prevent WWD.

Future research will attempt to find the most economical and cost-effective ways local agencies can coordinate to effectively respond to WWD events that occur on shared roadway systems or travel from one limited access roadway to another in the same county or across counties. Additionally, research can also be done on existing WWD countermeasures to determine if they are equally effective for drivers of all age and ethnic groups and for what specific reasons some groups (e.g., elderly or Hispanic) are overrepresented. WWD is a regional problem that requires multiple agencies to work together to detect, deter, and stop WW drivers, and to educate susceptible populations on the risks of WWD. With these regional efforts, highways can be made safer for all travelers.

Author Contributions

The authors confirm contribution to the paper as follows: study conception and design: Faruk, Al-Deek, Sandt, and Rogers; data collection: Faruk, Al-Deek, Sandt, and Rogers; analysis and interpretation of results: Faruk, Al-Deek, Sandt, and Rogers; draft manuscript preparation: Faruk, Al-Deek, Sandt, and Rogers. All authors reviewed the results and approved the final version of the manuscript.

References

1. National Transportation Safety Board (NTSB). *Highway Special Investigation Report: Wrong-way Driving*. Publication NTSB/SIR-12/01, Washington D.C., 2012.
2. Rogers, J., H. Al-Deek, A. Alomari, E. Gordin, and G. Carrick. Modeling the Risk of Wrong Way Driving on Freeways and Toll Roads. *Transportation Research Record: Journal of the Transportation Research Board*, 2016. 2554: 166–176.
3. Pour-Rouholamin, M., H. Zhou, B. Zhang, and R. E. Turochy. Comprehensive Analysis of Wrong-Way Driving Crashes on Alabama Interstates. *Transportation Research Record: Journal of the Transportation Research Board*, 2016. 2601: 50–58.
4. Rogers, J., A. Sandt, H. Al-Deek, A. Alomari, N. Uddin, E. Gordin, and G. Carrick. Wrong-Way Driving: Multifactor Risk-Based Model for Florida Interstates and Toll Facilities. *Transportation Research Record: Journal of the Transportation Research Board*, 2015. 2484: 119–128.
5. Sandt, A., H. Al-Deek, and J. Rogers. Identifying Wrong-Way Driving Hotspots by Modeling Crash Risk and

Assessing Durations of Wrong-Way Driving Events. *Transportation Research Record: Journal of the Transportation Research Board*, 2017. 2616: 58–68.
6. Rogers, J., H. Al-Deek, A. Alomari, F. Consoli, and A. Sandt. Wrong-Way Driving on Florida Toll Roads: An Investigation into Multiple Incident Parameters and Targeted Countermeasures for Reductions. *International Journal of Engineering Management and Economics (Special Issue on Transport)*, Vol. 5, No. 3/4, 2015, pp. 144–168. http://doi.org/10.1504/IJEME.2015.072549
7. Sicking, D., and K. Lechtenberg. Guidelines for Implementation of Cable Median Barrier. *Transportation Research Record: Journal of the Transportation Research Board*, 2009. 2120: 82–90.
8. Tamburri, T. Wrong-Way Driving Accidents Are Reduced. *Highway Research Record* #292, California Division of Highways, 1969.
9. Cooner, S. A., A. S. Cothron, and S. E. Ranft. *Countermeasures for Wrong-Way Movement on Freeways: Overview of Project Activities and Findings*. Report No. FHWA/TX-04/4128-1. Texas Transportation Institute, A&M University System, College Station, Texas, 2004. https://d2dtl5nnlpfr0r.cloudfront.net/tti.tamu.edu/documents/4128-1.pdf. Accessed July 2, 2017.
10. Cooner, S. A., and S. E. Ranft. Wrong-Way Driving on Freeways: Problems, Issues, and Countermeasures. Presented at 87th Annual Meeting of the Transportation Research Board, Washington, D.C., 2008.
11. Zhou, H., J. Zhao, R. Fries, M. R. Gahrooei, L. Wang, B. Vaughn, and B. Ayyalasomayajula. *Investigation of Contributing Factors Regarding Wrong-Way Driving on Freeways*. Report No. FHWA-ICT-12-010. Illinois Department of Transportation ICT-R27-90, 2012. https://www.ideals.illinois.edu/bitstream/handle/2142/45789/FHWA-ICT-12-010.pdf?sequence = 2&isAllowed = y. Accessed July 1, 2017.
12. Braam, A. C. *Wrong Way Crashes: Statewide Study of Wrong Way Crashes on Freeways in North Carolina*. Traffic Engineering and Safety Systems Branch, North Carolina Department of Transportation, 2006.
13. Haendeler, S., A. Lewandowski, and C. Wietfeld. Passive Detection of Wrong Way Drivers on Motorways Based on Low Power Wireless Communications. *Proc., IEEE 79th Vehicular Technology Conference (VTC Spring)*, IEEE, Seoul, South Korea, 2014, pp. 1–5. http://dx.doi.org/10.1109/VTCSpring.2014.7023039
14. Lathrop, S. L., T. B. Dick, and K. B. Nolte. Fatal Wrong Way Collisions on New Mexico's Interstate Highways, 1990–2004. *Journal of Forensic Sciences*, Vol. 55, No. 2, 2010, pp. 432–437. http://dx.doi.org/10.1111/j.1556-4029.2009.01305.x
15. Fariello, B. *San Antonio Wrong Way Driver Initiative*. Presentation. Texas Department of Transportation, 2012. https://conferences.tti.tamu.edu/traffic-safety12/program/9-breakout/fariello.pdf. Accessed July 1, 2017.
16. Finley, M., and N. Trout. Driver Expectations and Understanding of Wrong-Way Driver Warning Messages Displayed on Dynamic Message Signs. Presented at 96th Annual Meeting of the Transportation Research Board, Washington, D.C., 2017.
17. Sandt, A., H. Al-Deek, J. Rogers, and A. Alomari. Wrong-Way Driving Prevention: Incident Survey Results and Planned Countermeasure Implementation in Florida. *Transportation Research Record: Journal of the Transportation Research Board*, 2015. 2484: 99–109.
18. Federal Highway Administration [FHWA]. *Creating an Effective Program to Advance Transportation System Management and Operations*. Report No. FHWA-HOP-12-003. FHWA Office of Operations, U.S. Department of Transportation, Washington, D.C., January 2012.
19. Federal Highway Administration [FHWA]. *Getting More by Working Together: Opportunities for Linking Planning and Operations*. Report No. FHWA-HOP-05-016. FHWA Office of Operations, U.S. Department of Transportation, Washington, D.C., November 2004.
20. *Transportation Research Circular E-C166:* Glossary of Regional Transportation Systems Management and Operations Terms. Transportation Research Board of the National Academies, Washington, D.C., 2012.
21. *SHRP 2 Report 32 S2-L06-RR-2:* Guide to Improving Capability for Systems Operations and Management. Transportation Research Board of the National Academies, Washington D.C., 2011.
22. United States Census Bureau. *TIGER/Line® shapefiles and TIGER Line® Files*. https://www.census.gov/geo/maps-data/data/tiger-line.html. Accessed June 1, 2017.
23. Florida Highway Safety and Motor Vehicles. https://www.flhsmv.gov/resources/driver-and-vehicle-reports. Accessed June 5, 2017.
24. United States Census Bureau. *American FactFinder*. https://factfinder.census.gov/faces/nav/jsf/pages/index.xhtml. Accessed June 3, 2017.
25. *Traffic Safety Facts*. Publication NHTSA DOT HS 811 216. NHTSA, U.S. Department of Transportation, 2006. https://crashstats.nhtsa.dot.gov/Api/Public/ViewPublication/811216. Accessed June 10, 2017.
26. Correlli, E. C. *Crash Data and Rates for Age-Sex Groups of Drivers*. Research Note. NHTSA, U.S. Department of Transportation, 1998. https://www-nrd.nhtsa.dot.gov/pubs/98.010.pdf. Accessed June 10, 2017.
27. Lord, D., and F. Mannering. The Statistical Analysis of Crash-Frequency Data: A Review and Assessment of Methodological Alternatives. *Transportation Research Part A: Policy and Practice*, Vol. 44, No. 5, 2010, pp. 291–305. http://dx.doi.org/10.1016/j.tra.2010.02.001
28. SAS Institute Inc. *SAS/ACCESS® 9.4 Interface to ADABAS: Reference*. SAS Institute Inc., Cary, NC, 2013.

The Standing Committee on Regional Transportation Systems Management and Operations (AHB10) peer-reviewed this paper (18-01792) and selected it to receive its 2018 Best Paper Award.